What You Don't Know about Schools

Edited by

Shirley R. Steinberg
and
Joe L. Kincheloe

WHAT YOU DON'T KNOW ABOUT SCHOOLS
© Shirley R. Steinberg and Joe L. Kincheloe, 2006.

All rights reserved. No part of this book may be used or reproduced in any manner whatsoever without written permission except in the case of brief quotations embodied in critical articles or reviews.

First published in 2006 by
PALGRAVE MACMILLAN™
175 Fifth Avenue, New York, N.Y. 10010 and
Houndmills, Basingstoke, Hampshire, England RG21 6XS
Companies and representatives throughout the world.

PALGRAVE MACMILLAN is the global academic imprint of the Palgrave Macmillan division of St. Martin's Press, LLC and of Palgrave Macmillan Ltd. Macmillan® is a registered trademark in the United States, United Kingdom and other countries. Palgrave is a registered trademark in the European Union and other countries.

ISBN 1–4039–6344–4
ISBN 1–4039–6345–2 (pbk.)

Library of Congress Cataloging-in-Publication Data

    What you don't know about schools / edited by Shirley R. Steinberg and Joe L. Kincheloe.
      p. cm.
    Includes bibliographical references and index.
    ISBN 1–4039–6344–4—ISBN 1–4039–6345–2 (pbk.)
      1. Politics and education—United States. 2. Public schools—Political aspects—United States. I. Kincheloe, Joe L. II. Steinberg, Shirley R., 1952–

LC89.W46 2005
379.73—dc22   2005048685

Design by Newgen Imaging Systems (P) Ltd., Chennai, India.

First edition: January 2006

10 9 8 7 6 5 4 3 2 1

Printed in the United States of America.

Praise for Steinbe

S̶h̶ ...cape of
e̶d̶ ...ave done this through
t̶h̶ ...g publication series, through their
i̶n̶ ...borations with progressive educators around the
world, but most of all through their own brilliant and iconoclastic work.

> —Peter McLaren, Professor, University of California,
> Los Angeles, and author of *Capitalists and Conquerors*

Steinberg and Kincheloe provide readers with the very best of critical pedagogy and cultural studies.

> —Norman K. Denzin, University of Illinois at
> Urbana-Champaign

Steinberg and Kincheloe's work sparkles with the critical and creative energy of scholars and educators whose primary concern is with education that serves students on their own grounds, and helps our society realize the riches of diversity.

> —Kenneth J. Gergen, Mustin Professor of Psychology,
> Swarthmore College

Steinberg and Kincheloe are must reading for all of those whose ethics demand that they become agents of change in the difficult and constant struggle toward social justice and democracy.

> —Donaldo Macedo, Distinguished Professor of Liberal Arts and
> Education, University of Massachusetts, Boston

In celebration of the lives of Brenda Jenkins and Rosa Parks, who both died on October 24, 2005—both sisters in the struggle for racial justice and human freedom

Our beloved Brenda made us better for having known her

Sister Rosa ignited a spark of hope in a world of racial hatred

It is fitting that they left together

Contents

List of Abbreviations	vii
Notes on Contributors	viii

Chapter 1
What You Don't Know *Is* Hurting You and the Country   1
Joe L. Kincheloe

Chapter 2
How Did this Happen? The Right-Wing Politics of
Knowledge and Education   31
Joe L. Kincheloe

Chapter 3
What You Don't Know about Standards   69
Raymond A. Horn, Jr.

Chapter 4
What You Don't Know about Evaluation   89
Philip M. Anderson and Judith P. Summerfield

Chapter 5
The Cult of Prescription—Or, A Student Ain't No
Slobbering Dog   103
P. L. Thomas

Chapter 6
School Leaders, Marketers, Spin Doctors, or Military
Recruiters?: Educational Administration in the
New Economy   115
Gary L. Anderson

Chapter 7
The Price for "Free" Market Capitalism in Public
Schools—or How much is Democracy Worth on the
Open Market?   127
John Weaver

Chapter 8
Bad News for Kids: Where Schools Get Their News for Kids   141
Carl Bybee

Chapter 9
Meritocratic Mythology: Constructing Success 169
Benjamin Enoma

Chapter 10
What Is Not Known about Genius 183
Ray McDermott

Chapter 11
What You Don't Know about Diversity 211
Elizabeth Quintero

Chapter 12
Hiding in the Bathroom: The Educational Struggle of Marginalized Students 223
Danny Walsh

Chapter 13
When Ignorance and Deceit Come to Town: Preparing Yourself for the English-Only Movement's Assault on Your Public Schools 241
Pepi Leistyna

Chapter 14
Live from Hell's Kitchen, NYC 261
Regina Andrea Bernard

Chapter 15
Letting Them Eat Cake: What Else Don't We Know about Schools? 277
Shirley R. Steinberg

Index 286

List of Abbreviations

ABCTE	American Board for Certification of Teacher Education
AES	American Eugenics Society
CEO	Chief Executive Officer
ELLs	English Language Learners
ETS	Educational testing Service
GAO	General Accounting Office
GMAT	Graduate Management Admissions Test
GRE	Graduate Record Examination
ISLLC	Interstate School Leaders Licensure Consortium
JROTC	Junior Reserve Officers Training Corps
LAD	Language Acquisition Devise
LEP	Limited English Proficient
LSAT	Law School Admissions Test
MCAS	Massachusetts Comprehensive Assessment System
MCAT	Medical College Admissions Test
NAEP	National Assessment of Educational Progress
NCATE	National Council for Accreditation of Teacher Education
NCLB	No Child Left Behind
NCLR	National Council of La Raza
NIE	Newspapers in Education
NRC	National Research Council
NRP	National Reading Panel
SAT	Scholastic Aptitude Test
SES	Socioeconomic Status
TQM	Total Quality Management
TVA	Tennessee Valley Authority
WRC	Weekly Reader Corporation
ZPD	Zone of Proximal Development

Notes on Contributors

Gary L. Anderson is a professor in the Steinhardt School of Education, New York University. His research interests include educational leadership, critical ethnography, action research, and the impact of neoliberalism on schools. Recent books include *Performance Theories and Education: Power, Pedagogy, and the Politics of Identity* (coedited with Bryant Alexander and Bernardo Gallegos), and *The Action Research Dissertation: A Guide for Students and Faculty* (coauthored with Kathryn Herr).

Philip M. Anderson is Professor of Secondary Education at Queens College, and Professor of Urban Education and Executive Officer of the PhD Program in Urban Education at the City University of New York (CUNY) Graduate Center. He studies culture and curriculum in schooling, in particular, as it relates to aesthetics and humanities education.

Regina A. Bernard was born and raised in Hell's Kitchen. She received her BS in Criminal Justice from John Jay College. In 2001, she was the first student to graduate from Columbia University's MA program in African American Studies. She will be the first graduate of the CUNY Urban Education PhD program, completing her dissertation on Critical Race Study.

Carl Bybee is the director of the Oregon Media Literacy Project and an Associate Professor of Communication Study in the School of Journalism and Communication at the University of Oregon. His current research focus is on helping youth find a new vocabulary for becoming engaged democratic citizens drawing on the connections between news, entertainment, media literacy, and new models of active democracy.

Benjamin Enoma codirects the recruitment and enrollment services for the honors, flextime, and accelerated part-time MBA programs at the Zicklin School of Business, Baruch College. His area of research is educational policy.

Raymond A. Horn, Jr. is an assistant professor of education at Saint Joseph's University and the director of the Interdisciplinary Doctor of Education Program for Educational Leaders. He is the author of *Teacher Talk: A Post-Formal Inquiry into Educational Change*.

Joe L. Kincheloe is the Canada Research Chair at the McGill University Faculty of Education. He is the author of numerous books and articles about pedagogy, education and social justice, racism, class bias, and sexism, issues of cognition and cultural context, and educational reform. His books include *Teachers as Researchers, Classroom Teaching: An Introduction, Getting Beyond the Facts: Teaching Social Studies/Social Sciences in the Twenty-First Century; The Sign of the Burger: McDonald's and the Culture of Power; City Kids: Understanding Them, Appreciating Them, and Teaching Them, and Changing Multiculturalism* (with Shirley Steinberg). His coedited works include *White Reign: Deploying Whiteness in America* (with Shirley Steinberg et al.) and the Gustavus Myers Human Rights award winner *Measured Lies: The Bell Curve Examined* (with Shirley Steinberg). Along with his partner, Shirley Steinberg, Kincheloe is an international speaker and lead singer/keyboard player of Tony and the Hegemones.

Pepi Leistyna is an associate professor in the Applied Linguistics Graduate Studies Program at the University of Massachusetts, Boston. His books include *Breaking Free: The Transformative Power of Critical Pedagogy; Presence of Mind: Education and the Politics of Deception; Defining and Designing Multiculturalism; Corpus Analysis: Language Structure and Language Use*; and *Cultural Studies: From Theory to Action*.

Ray McDermott is a professor of education at the Stanford University. He is interested in the analysis of human communication, the organization of school success and failure, the history of literacy and urban education. He has published numerous books and articles in these areas. In 2001 he won the Council on Anthropology in Education Spindler Award for anthropological research in education.

Elizabeth Quintero is Associate Professor in the Department of Teaching and Learning at NYU. Her teaching, research, and service involve working with multilingual families, students, and young children. She uses critical literacy focusing on family stories and multicultural literature to support students' strengths and their journeys toward school success. She is the coauthor of *Becoming a Teacher in* the New Society: Bringing Communities and Classrooms Together (with Mary Kay Rummel) and the author of Problem-Posing with Multicultural Children's Literature: Developing Critical, Early Childhood Curricula.

Shirley R. Steinberg is an associate professor at McGill University Faculty of Education. She is the author and editor of numerous books and articles and coedits several book series. She is also the founding editor of *Taboo: The Journal of Culture and Education*. Steinberg has recently finished editing *Teen Life in Europe*, and with Priya Parmar and Birgit Richard *The Encyclopedia of Contemporary Youth Culture*. She is the editor of *Multi/Intercultural Conversations: A Reader*. With Joe Kincheloe she has edited *Kinderculture: The Corporate Construction of Childhood* and *The Miseducation of the West: How Schools and the Media Distort Our Understanding of the Islamic World*. She is coauthor of *Changing Multiculturalism: New Times, New Curriculum*, and *Contextualizing Teaching* (with Joe Kincheloe). Her areas of expertise and research are in critical media literacy, social drama, and youth studies. She sends a shout-out to all her colleagues and friends who have suffered the absurdity of U.S. national "accreditation" these past years.

Judith P. Summerfield works at the CUNY as University Dean for Undergraduate Education, Professor of English at Queens College, and Professor of Urban Education in the PhD Program at the Graduate Center. Named the New York State Professor of the Year by the Carnegie Foundation for the Advancement of Teaching in 1998, her current research focuses on undergraduate education in urban contexts.

P. L. Thomas is an assistant professor of Education at Furman University in South Carolina. He is the author of *Numbers Games, Teaching Writing Primer*, and *Reading, Learning, Teaching Barbara Kingsolver*.

Danny Walsh teaches English for the ninth and tenth grade in a new, small high school in Brooklyn that is designed to serve recent immigrants. He is finishing a PhD in urban education at the CUNY Graduate Center.

John A. Weaver is an associate professor of Curriculum Studies at Georgia Southern University. He is the author or coeditor of *Popular Culture and Critical Pedagogy, (Post) Modern Science (Education), and Science Fiction Curriculum, Cyborg Teachers, and Youth Cultures*.

Chapter 1

What You Don't Know *Is* Hurting You and the Country

Joe L. Kincheloe

As a writer I strive to avoid overstatement—but in the middle of the first decade of the twenty-first century, discussions about education lend themselves to dramatic proclamations. Take this one, for example: public education in the United States is facing the greatest threat to its continued existence in its 150-year history. Right-wing groups enjoying more power than ever before are constructing a crisis in education that can be solved, they argue, only by ridding ourselves of so-called government schools. Because public education is deemed to be "beyond hope," only the creation of a system of corporate-run private schools can assure quality education. As right-wing think tanks and conservative composed governmental reports manipulate data to indicate that thousands of schools are failing, "qualified" teachers are in perilously short supply, and schools are indoctrinating students with "radical" ideas, Americans are being convinced that government-supported schools cannot deliver what the nation needs. Most Americans do not know about such tactics and if such public unawareness continues, we could soon witness the end of universal public schooling for our young people.

Cooking the Books: Creating Chaos

The climate of deceit that has been created produces misinformation and erases knowledge that might contradict such deception. For example, the No Child Left Behind (NCLB) federal law's mandated "adequate yearly progress" formulas—how we measure the performance of schools—are so unclear and blurred that it seems to many observers that their real goal is the construction of confusion and failure. Under these accountability guidelines as many as three out of four schools will be deemed to not be making sufficient progress. Included in this group will be some of the best public schools in the country. Such schools will be publicly embarrassed by government censure of their inadequate performance, leading, hopefully, more people to conclude that public schools are failing.

In the spirit of this censure and the climate of fear it produces, right-wing promoted voucher plans have been passed by state legislatures that provide parents with minimal tuition funds so they can take students out of "failing" schools and enroll them in private or religious schools. Voucher supporters know that this removes money from public school funds, thus making it more difficult for them to operate. The point in many of these plans is not to improve education but to create bad press for and punish public schools. After its capture by conservative groups, the federal Department of Education has covertly operated to reward organizations that are willing to provide support for the right-wing privatization effort. Operating with a preordained ideological agenda, these groups produce data that "proves" that even schools such as Virginia's Langley High in Fairfax County— more than 90 percent of its graduates go to college—are failing (Weil, 2001; Karp, 2002; Metcalf, 2002; Aratani, 2004; WEAC, 2004).

In spirit of the political strategizing of Karl Rove and the late Lee Atwater, right-wing operatives understand the partisan value of privatization—or school choice as it is labeled. Education has traditionally been an issue that helps the political left—in the name of opportunity, mobility, and the promotion of democratic values, left-leaning politicians have promoted public education. If the public could be convinced that most public schools are failing, conservatives reasoned, then promises to fund public education would begin to ring hollow. School choice could be promoted to the poor and the racially marginalized as an issue of justice and equality. Along with religion, the privatization of education could be employed to subvert traditional progressive political constituencies. In this context, school choice has been deceptively promoted as the new front of the twenty-first century civil rights movement. As the conservative Alliance for the Separation of School and State puts it, we must end "government involvement in education" (Miner, 2004).

In many ways, the privatization of public schooling—the end of public education—is the big jackpot for the right. Such a victory, conservatives reason, will subvert one of the last mechanisms for possibly producing knowledge that questions or conflicts with right-wing disinformation. Admittedly, the willingness and ability of public education to counter right-wing politics of knowledge over the last few decades have declined precipitously. With the destruction of public schools, however, the "threat" of a space that offers a diversity of knowledges and ideologies profoundly recedes. The language of rigor, standards, and accountability resonates with the public, and too few individuals have looked behind the words to the actual policies implemented in the name of such concepts. In addition to dumbing down schools by requiring them to teach to standardized tests emphasizing the memorization of meaningless, fragmented, and decontextualized data, such policies exacerbate the gap between the rich and the poor and white and nonwhite in U.S. society (Berkowitz, 2001; Hartman, 2002).

Ideological Assumptions behind Right-Wing Education

Ideology is traditionally defined as a system of beliefs, but I use the term in this context to denote something a little more complex. Dominant ideological activity in the context of critical theory involves the process of protecting unequal power relations between different groups and individuals in society. For example, dominant ideology sustains unequal power relations via the process of making meaning—in a sense by "educating" and "reeducating" the public. Thus, the way I use ideology here involves the progressive concern with the way oppression takes place and the power disparity that accompanies it. Applying the concept of ideology as we explore what many people do not know about education in the current mediascape and electronic information environment, we begin to understand how powerful groups shape people's consciousness in ways that will better serve the interests of dominant power. In such an ideological environment with its corporate-backed power to persuade people of the worth of the privatization agenda, we can begin to see the interests such a policy would serve—and not serve. With privatization, for example, opportunities would be created for new ways for business to make billions of dollars of profit from for-profit schools and the child consumers who attend them. Right-wing ideology makes such capital production possible.

It is easier to wield power in a privatized society than in a public one. Indeed, privatized power is accountable to very few, as most citizens have no say over who does what in a corporate-run institution. Right-wing ideology has successfully produced a political climate where millions of people have come to believe that public ownership of social organizations is a manifestation of oppression whereas private ownership is the ultimate marker of freedom. Interestingly, such representations of public ownership have been more successful in recent decades than during the cold war when old conservatives equated public ownership of institutions, such as the Tennessee Valley Authority (TVA), to communism. Questions concerning accountability of private organizations have been adeptly swept under the rug in such ideological representations.

In the United States, such a right-wing ideological success has helped to usher in a new political era. Some might call this success an ideological reeducation of the American public. In this new political era, the demands of the market always trump the needs of the larger society as well as the perpetuation of democracy and democratic institutions. In the new cosmos, the government no longer intervenes to promote equity and protect the needs of those treated unfairly because of race, class, or gender. Indeed, in the brave new world of education, traditional conservative values, such as local control of schools, collapse in the face of dominant ideology of privatization. The federal government's role in twenty-first century education is to protect market needs by promoting a national agenda of standardization and privatization.

Thus, when President George W. Bush signed NCLB into law on January 8, 2002, the federal government's role in K-12 schooling shifted from equity to guaranteeing simplistic and reductionistic forms of accountability. Such an assertion should not be taken as a rejection of school accountability; instead, it is an assertion that the *types* of accountability mandated often reconstruct school purpose in a way that promotes low-level thinking skills and reduces education to the indoctrination of unchallenged "truths." The law's claim of increased flexibility and local control is misleading doublespeak, and its focus on teaching strategies that have been scientifically proven to improve instruction raises profound issues about the nature of knowledge production in a democratic society.

As for equity, the law does nothing to address the grotesquely unequal funding that separates schools and school districts in well-to-do and poor neighborhoods. In a public school system that gives lip service to an ideology of equality, the neglect of such equity issues shackles the progress of poor schools and the students who attend them. Yet, the ideological refraction promoted by the right-wing rhetoric frames these ideological-driven reforms as the educational salvation of the dispossessed. Indeed, contemporary educational politics is best understood as a campaign of class disinformation. No wonder so many Americans know so little about these issues.

The doublespeak is pervasive in the disinformation campaign. Under the banner of "getting big government off our backs," the right-wing education program mandates new forms of control over what can and cannot be taught in schools. Simultaneously, it unleashes a socially unregulated corporate power that oppresses the lives of people in new and more insidious ways. The combination of the two mechanisms of regulatory power places democracy and individual freedom at risk. Existing government structures regulate education, while working in the long run for corporate ownership of schooling. In this ideological context, the U.S. General Accounting Office (GAO) found that the corporate commercialization of public schooling is expanding rapidly. In most public schools, one can now find sale of corporate products, corporate advertising, and corporate marketing studies of student consumption patterns. Through these intrusive and exploitative activities, corporations have generated 150 billion dollars (and growing) in annual sales to teenagers alone (Hursh, 2001; Hartman, 2002; Foley and Voithofer, 2003; Steinberg and Kincheloe, 2004).

All of these activities are justified under the attack on the public domain. As a traditional American marker of the public space, education was first on the right-wing hit list. In the right-wing politics of information of the twenty-first century with its talk radio, Fox News Channel, and increasing corporate control of other broadcast and print media, education is a central target but only one of many. Publicly supported programs that address issues of public welfare and support for the least affluent members of society, such as Social Security, Medicare and Medicaid, and college aid to students in need, have also been under attack. In the right-wing ideology of disinformation in education, for example, public schools are consistently positioned as being inferior to private schools. Private schools take the same types of students that attend public schools, antipublic education operatives proclaim, and for less money turn out better educated graduates. I have rarely read in one of these private-beats-public studies that public schools are required by law to admit all students who apply whereas private schools can choose whom to admit. Many scholars (Bogle, 2003) of privatization conclude that despite this advantage for private institutions, the difference in the performance of public and private school students is negligible.

Following its narrow ideological agenda, the Bush administration has sought to undermine public education one piece at a time—always, of course, in the name of improving it. One of the major fronts of attack has been directed toward public school teachers. Over the last few years, antipublic education groups had worked for the deprofessionalization of teachers. Such a movement had made little progress until the election of George W. Bush in 2000. As part of his larger plan for public education, Bush funded these groups with millions of dollars, with new support at the highest levels and funding proponents of deprofessionalization formed the American Board for Certification of Teacher Education (ABCTE) to promote a simplistic form of teacher certification characterized by few requirements. At the same time it claims that schools and teachers are failing and need higher standards to promote "educational excellence," the right throws its full weight behind efforts to undermine high standards in the professional preparation and certification of teachers (WEAC, 2004). Such a schizophrenic scheme fits well the devise standards, reduce funding and support, test and measure, proclaim failure, then privatize grand strategy.

Once this privatization achieves success, the profound class differences that separate Americans in the middle of the first decade of the twenty-first century will expand dramatically. Access to education for the poor and racially marginalized will become harder and harder, for there will be few incentives for private for-profit schools to admit such students. A largely privileged, white corps of students will gain better access to the scientific, technological, and information professions, while the poor and the minority young people will be left with low-pay, low-benefits service sector jobs. Such a bimodal distribution of privileged and marginalized workers will not only cause suffering for the marginalized but will also place great stresses on the American social fabric in the coming years. Such an inegalitarian future is dystopic as well as contrary to the social compact that the nation has at least given lip service to in the past.

The ideological foundation on which right-wing education rests has been articulated clearly by conservative think tanks. For example, in the Heritage Foundation's education manifesto, *No Excuses: Lessons from 21 High-Performing, High-Poverty Schools*, the continuation of high poverty rates is not viewed as a problem with which we should be concerned. In addition to promoting tax cuts for the most wealthy members of society, drastic reduction in spending for social programs, and the end of the minimum wage, and other measures that exacerbate the growth of poverty, the Heritage Foundation claims that the poor are generally immoral and criminally inclined, that is, not worth helping in the first place. In fact, spokespeople for the institute continue, the so-called poor are not as bad off as many would have us think. Many of them own numerous cars, expensive kitchen appliances, and hot tubs (Coles, 2003). The logic of the welfare Cadillac is alive and well at one (and there are more) of the leading scholarly institutes in the country. In this right-wing ideology, poverty is the fault of the poor, and as such why should we (white, upper-middle class Americans) worry about educating the next generation of their spawn?

The Ideological Foundation: Crass Class Politics and Corporate Rule

Such a class-biased ideology is nothing new in the history of American education. The greatest difference is that it is being carried out so insidiously as part of a campaign to improve education for everyone especially the poor and the racially marginalized. The goals of this twenty-first-century stealth campaign are little different from, say, nineteenth- and early-twentieth-century efforts to discipline an industrial workforce that would work in low-skill and boring factory jobs. Then, as now, corporate and business leaders would have been reluctant to support any educational plan that did not result in monetary profit for themselves and other dominant power groups. Because of the control of school boards and state departments of education by the elites, there was little chance throughout the history of American education for school policy to rarely run counter to the best interests of dominant groups. To understand twenty-first-century tactics, analysts must view them in the context of the right-wing ideological goals of the last 30 years.

In 1980, for example, the average business or corporate chief executive officer (CEO) earned 38 times the salary of an average schoolteacher and 42 times as much as an average factory worker. By 1990, after ten years of right-wing economic redistribution policies, the average CEO earned 72 times as much as a teacher and 93 times as much as a factory worker. In 2004, the average CEO earned more than 500 times than the average factory worker and more than 1,200 times than a worker making the minimum wage (Coontz, 1992; Gonzalez, 2004). Crass class politics continues with little resistance and is changing the social landscape of the country. Americans have always placed great value on hard work. People who work hard, it is commonly believed, should be rewarded for their efforts. Most Americans would be surprised to find out, therefore, that the redistribution of wealth over the last 25 years has been accomplished in inverse relation to hard work.

Much of the wealth created in the contemporary United States did not come from inventing a better mousetrap or long hours of study or working overtime. Most new wealth befell those with enormous assets who were able to reap "instant wealth" from rapidly fluctuating return rates on their speculative investments. Dividends, tax shelters, interest, and capital gains were at the center of the action—not hard work. The right-wing ideological assertion that connects one's class position to one's willingness to work hard may be less direct than many Americans have assumed. The right's attempt to dismiss class as an American issue must be exposed for what it is—an instrumental fiction designed to facilitate the perpetuation of the growing disparity of wealth by pointing to the laziness and incompetence of the poor as the cause of their poverty. When progressives raise class issues in relation to the educational policies promoted by conservatives in the last three decades of the twentieth and the first decade of the twenty-first centuries, it is consistently dismissed as an effort to excuse weak teachers and failing students. Educational proposals such as the hidden movement for corporate-run private schools and the deprofessionalization of teachers must be viewed in larger historical and class warfare contexts. They contribute to a long-term right-wing effort to further marginalize the poor and extend the privilege of the well to do.

Technology and mass communication corporations in the middle of the twenty-first century exercise unprecedented power to control information. Because of their control of more and more print and broadcast media outlets, corporate power has never been more entrenched. Recognizing that teachers and the public institution of education could pose a threat to such an information monopoly, many corporate leaders have led the privatization and teacher deprofessionalization movements. Thus, education has been swept up in the same ideological forces that have redefined freedom as the right of corporations to desecrate the public space in an effort to pursue private gain. Data banks, radio and television transmissions, Internet, and other forms of transnational communications systems all contribute to a network that allows corporate leaders to regulate markets and manipulate public opinion all over the world.

As these communications systems filter into cities, villages, and rural areas globally, corporations present a view of the world that promotes their interests. This privatized educational revolution and the changes in ideological consciousness it produces takes place below the radar of public awareness. It is rarely studied in elementary or secondary schools and is even hard to find in the curriculum of higher education. It can be found, thankfully, in a few media literacy classes in teacher education and in media studies departments in communications. With little fanfare, it implicitly promotes values such as competitive individualism, the superiority of an unregulated market economy, individual blame for poverty, the irrelevance of moral and philosophical types of knowledge, and the necessity of consumption as part of a larger quest for social status. People's identities—their sense of who they are—begin to be formed less and less in their communities and personal interactions and more and more by their televisions and other corporatized information sources.

This corporatized politics of information profoundly affects an individual's perception of the world. This, many argue, is in the twenty-first century central to an understanding of the contemporary state of democracy. While it is hard to spot in increasingly corporatized public schools, it is impossible to find in for-profit corporate schools. Teaching about the politics of information in such settings would violate the prime directive in corporatized pedagogy: do not call attention to the existence of power and the way it is wielded. In this context, we begin to discern that the corporate control of schools operated by deskilled, minimally educated teachers fits nicely with the larger effort to control public information and regulate public consciousness.

Make no mistake, the shaping of public opinion by way of corporate media and educational control is never simplistic and uncontested. Often, efforts to manipulate opinion backfire, as people begin to perceive what is happening to them and rebel. Furthermore, technologies such as computers and the Internet can be used to convey alternative messages that challenge corporate control. Still, most individuals in the middle of the twenty-first century do not comprehend the degree of influence that corporate leaders attain as they control television and other media that bypass reason and focus directly on the management of human feelings and emotions. Indeed, some of the most important ideological tools in the politics of information involve media presentations that are not overtly political.

Images of children as they open gifts on Christmas morning, for example, have no overt political message. At a deeper level, however, such images may be influential as they tell us that such happiness in our children can be evoked only by the consumption of goods and services. And where do we get these things? From businesses and corporations. If we truly love our children and want to see them happy, then we must support the interests of the companies that provide these valuable products. The production of ideological consciousness is not a linear, rational procedure but one grounded on our emotional hopes and fears. Thus, when Mattel Inc. calls for lower corporate taxes and a better business climate in which to produce its toys, we accede to its wishes. After all, this is the company that allows us to make our children happy (Kincheloe and Steinberg, 2004; Steinberg and Kincheloe, 2004).

Thus, it is within the context of this crass class politics and corporate power that we begin to better understand the right-wing educational agenda. Just as corporations and conservative groups have worked to control the politics of information in communications, the same forces are at work

1. to regulate the subject matter of schools in the most efficient way possible;

2. to control the work and academic freedom of teachers;
3. to subvert the possibility that a variety of knowledges might be engaged in public spaces.

Right-wing advocacy of basic skills teaching accompanied by multiple-choice standardized testing viewed in this context not only tends to dumb down the curriculum, but it also keeps teachers and students from exploring dangerous information—for example, the problems of democracy as John Dewey put it so long ago. When the Heritage Foundation encourages educational leaders to employ the "hiring and firing of staff" to let them know what it is that a school is supposed to accomplish, one senses how serious such political operatives take their regulatory task.

What schools, especially high poverty schools, are supposed to accomplish involves preparing a student to fit the needs of a competitive workplace. Careful reading of what this means reveals that fitting such needs involves learning to follow directions, respect authority, eschew unions, and ask fewer questions. Such forms of oppressive schooling are part of a larger corporate strategy to disempower and control workers. Over the last 30 years, corporations have developed new forms of worker surveillance to ensure that laborers are always under supervision. Such surveillance includes computer tracking of practically every move a worker makes. Tens of millions of people are hired to police workers—about one monitor for every 2.3 workers. Indeed, a key dimension of what schools should accomplish, according to the Heritage Foundation and other right-wing groups, involves the same forms of control. Rarely will students study the antisocial behavior of corporations in the standardized curricula of right-wing schools. Students will not be exposed to the growing disparity of wealth between workers and managers. Indeed, much of their education will involve simple-minded homilies promoting a corporation-is-our-friend ideology (Coles, 2003; Kitts, 2004).

Science as Right-Wing Ideology: Understanding Positivism

A central and disturbing development in the right-wing educational movement involves the deployment of science in circumscribed ways guaranteed to validate the power-driven dominant cultural agenda. This strategy is quite complex and it is easy to exclude the public from an understanding of how it works. The use of such a science for anti-democratic and antiegalitarian objectives has become extremely important in the twenty-first century. In this context, it is imperative that the public understands this politics of knowledge in order to defend democracy and access to public education. One of these complex elements involves the ability to identify and trace the effects of ethnocentrism within the positivist research tradition in education. Positivism is an epistemological (having to do with the production of knowledge) position that values objective, scientific knowledge produced in rigorous adherence to the scientific method. In this context, knowledge is worthwhile to the extent that it describes objective data that reflect the world. Over the past several decades, scholars of research have discerned numerous problems with the positivist position—problems that lead to the production of very misleading understandings of the world around us.

In this positivist view, "true knowledge" can only be produced by a detached, disinterested, external observer who works to ignore background (contextual) information by developing "objective" research techniques. In the long course of human history, most of great wisdom has not been constructed in this manner. At the center of the things we do not know about the right-wing movement in education is this attempt to recover and reinstate the positivist mode of producing a narrow form of knowledge. A more global insight with awareness of and respect for diverse ways of knowing, cultural humility, and an ecologically sustainable and ethical conception of progress is not on the positivist conceptual map. Positivism in this context is a monocultural way of seeing the world that emphasizes the knowledges produced by patriarchy, white Europeans, and individuals from the upper middle/upper classes. The standards movement of the last couple of decades provides a case study of this phenomenon.

In 1994, when Lynne Cheney (wife of Vice-President Dick Cheney) was attacking the National History Standards from her post at the conservative think tank, the American Enterprise Institute, she objected to their excessive coverage of women and minorities. The professional historians, she argued, who wrote the "disastrous" standards were anti-American radicals out to destroy Western civilization and the Enlightenment tradition. Even the most minor attempts to include diverse voices in the history curriculum in the U.S. schools are met with vicious objections. What is especially amazing in this situation is that the National History Standards was not calling for a major overhaul of curriculum to include the study of global, non-Western, and non-Christian information in the history curriculum in U.S. schools. The call, by scholars such as myself and the authors included in this book, for diverse global understanding, respect for the traditions of other cultures, and ways of producing knowledge are dismissed by the right as an assault on "all we hold dear." It is essential that progressives understand this arrogance, its numerous consequences around the world and within Western societies, and develop the skills to counter its expression and negate its unfortunate consequences (Apple, 1993).

In this context, the call for positivistic research and curricular standardization takes on even more ideological baggage. Not only a manifestation of hyper-rationalization, but also the standardization of curriculum becomes a means of insuring ethnocentrism in the classroom. Such an ethnocentrism is suspicious of concepts such as diversity, multiple perspectives derived from multiple forms of research, criticality, difference, and multiculturalism. Ideologically, it works covertly to promote the interests of dominant culture over less powerful minority cultures. Such interests involve the power of the privileged to maintain their privilege, as students from economically poorer families, those students whose families possess the least formal education, are transformed into "test liabilities" (Ohanian, 1999; Vinson and Ross, 2001). In such a category, their problems in school can be blamed on their inferiority: "we tried to teach them the information mandated by the standards but they just didn't have the ability to get it. There's nothing more we can do. Scientific analysis shows these students just can't learn. Look at their IQ scores."

Positivism exerts a dramatic impact on education. It is grounded on the faith that education like the physical and social worlds is founded on universal and unchanging laws. Thus, the purpose of educational research is to discover those laws. Educational laws would include statements about how students learn and how they should be taught. Positivist research tells us that there is one right way to teach and one correct way to evaluate that teaching. Those who hold different perspectives on these matters do not fit in the positivist universe. The evidence-based research promoted by the Bush administration and right-wing organizations fits the following categories of positivism:

(1) All legitimate knowledge is scientific knowledge—all scientific knowledge is empirically verifiable. Empirically verified knowledge is the information we gain through the senses. What the eye sees, what the ear hears, what we can count, what we can express mathematically—these constitute empirical knowledge. Of course, what I argue here is that many dimensions of education, psychology, and social activity in general are not discernible in such an empirical context. What is the purpose of schools in a democratic society is not a question that can be expressed empirically in the language of mathematics. Many normative and affective dimensions of education cannot be delineated in such a positivistic discourse. When the public encounters

such positivistic knowledge—for example, school progress expressed solely in test scores—it must understand that it presents a very limited view of the teaching and learning process. Indeed, some of the most important dimensions of education involving questions of equity, fairness, justice, and even intellectual rigor are erased from such forms of knowledge production. Knowledge about the world, and about the educational domain in particular, is never objective and disinterested. It is always based on a constellation of values and assumptions about the ways the world operates and the nature of human beings. These values and assumptions covertly shape the knowledge even an "objective" researcher produces.

(2) Researchers must use the same research methods to study the world of education that they use to study the physical universe. Profound problems emerge when researchers apply physical science methods to the study of the social world or education. A key characteristic of positivistic research in the physical sciences involves the effort to predict and control natural phenomena. When employed in education, positivistic physical science methods deploy knowledge as a tool to control people. Thus, ability testing is used to control where particular students can go, what programs they can pursue. Many a student of color or an economically marginalized student has heard a guidance counselor say on the basis of a positivistic aptitude test: "I'm sorry, Jamie, you will have to go to the basic classes. The tests tell us that you are not upper-track or college material."

(3) The world is uniform and the object of study will always be consistent in its existence and its behavior. Positivist researchers believe that the phenomena they examine will remain constant over time. Thus, they assume there is a natural order in the way both the physical and the social/educational/psychological worlds operate. These regularities, or social laws, are best expressed through quantitative analysis using propositional language and mathematics. The goal for such evidence-based educational research in this context is to develop theories that regularize human expression and make it predictable. In this context, right-wing groups study which curricula and pedagogies help raise test scores. If a particular teaching technique works in one locale, then because of the alleged consistency across educational contexts, it will exist in another. Following such positivist logic, educators should standardize their teaching in line with such research, and all schools should operate in the same way. Of course, this is exactly what NCLB is mandating in U.S. schools in the middle of the first decade of the twenty-first century.

(4) The variables that cause things to take place are limited and knowable, and, in positivist educational research, can be controlled. Positivists

assert that causal variables can be isolated and analyzed independently to determine specific cause–effect relationships. Thus, positivist research produces knowledge that is certain that direct inculcation of data by the teachers directly leads to better learning. Of course, in such a context, the ambiguity and complexity of what we might mean by the words "better learning" is simply ignored. For the cause–effect link to work in such research, we must not question the idea that student compilation of numerous decontextualized and fragmented "facts" is tantamount to being an educated person. In order to devise studies that claim to discover cause–effect relationships in education, researchers must reduce the factors studied to the point that the data produced is meaningless. Thus, they ignore hundreds of variables concerning particular value assumptions and the "noise" of everyday classroom life—the very dynamics that shape and give meaning to what education is all about.

(5) *If researchers use rigorous quantitative methods carefully then we can be certain about what is required in the reform of education—eventually we will understand education well enough to preclude the need for further educational research.* Educational researchers who are more aware of the different types of knowledge needed in educational research, and the complexity of contextual difference with diverse student needs, know that we can never control all the variables in an educational situation. Because the factors that help shape teacher and student behaviors are unlimited, the quest for positivist certainty is fatuous. Our understandings of the world in general and education in particular change with new revelations and they will continue to change. It is a naïve understanding of knowledge that assumes educational research will someday become unnecessary because we know all the answers. One of the great failures of the mindset that promotes right-wing educational reforms involves the arrogant assumption that we know how to teach and we must impose this knowledge on all teachers in all schools. In this framework, there is no room for teacher prerogative and individual innovation.

(6) *Objectivity is possible—facts and values in research about education can always be kept separate.* Evidence-based research is never a value-free activity. When positivist researchers assume, for example, that standardized test scores provide a valid measure of the quality of teaching and learning, they are accepting on faith a particular value. Test scores do not measure the ability to think at a higher order, make wise decisions, conduct and evaluate research, construct compelling interpretations of disparate information, discern validity, ad infinitum. I would argue that all of these abilities—another value judgment—are central qualities possessed by educated people. When teachers are

forced to teach to the fragmented and decontextualized demands of standardized tests, we observe a research instrument—the tests—shaping the nature of the teaching and learning that takes place in the classroom. The tail is wagging the dog.

(7) *There is only one true reality—the one "discovered" by positivist research—and the purpose of education consists of teachers conveying that reality to students.* Educational science grounded on this positivist tenet assumes that the laws of society and knowledge about human beings are certified and unchanging and, thus, ought to be inserted directly into the minds of students. Operating on this assumption, educational "engineers" devise curricula and teaching strategies for schools as if no ambiguities or uncertainties in the socioeducational world exist. The authoritarian voice of positivist educational science silences our language of qualitative insight based on experience, aesthetics, historical context, interpretive analysis, and descriptive processes. In such a context, a distanced, decontextualized, de-emotionalized, dehumanized, simplistic, and highly misleading form of knowledge is produced.

(8) *Devalues the professional complexity of the teaching act—teachers become information deliverers not highly skilled and respected scholar practitioners.* Here we can clearly discern the resonance between right-wing educational proposals and the positivist politics of knowledge. A positivist mode of educational research propels the deprofessionalization agenda by making teacher deskilling technically necessary. If teachers are merely delivering certain truths produced by experts, then there is no need for a scholarly teacher corps. Indeed, we can hire individuals who read on about the eighth or ninth grade level to read scripted lessons to students. Such "teachers" should be large and physically intimidating individuals who can frighten students into staying on task. Of course, in the corporate, for-profit schools referenced earlier in this chapter, these are exactly the teachers for whom school leaders are looking. The scholarly, knowledge-producing teachers advocated here are often viewed as undesirables, potential troublemakers in the right-wing school of the contemporary era.

A Science of Dominant Power Accompanied by Corporate Cronyism

In 2002, NCLB specifically endorsed an educational research limited to evidence-based scientific methods—positivist science—and insisted that only teaching strategies "proven to work" by such methods be used in schools. This "recovery" of a positivism that had been discredited by numerous scholars over the past three decades is part of a larger cultural movement. This global movement has attempted to turn back a tide of anti-European colonialism, garner respect for diverse cultural ways of understanding the world, promote a struggle against patriarchy, and initiate a quest for multiple-research methods that dig deeper into social, psychological, and educational realities. Drawing upon the work of Aaron Gresson (1995, 2004), I have referred to this "backlash" as the recovery movement. Such an effort has typically involved Westerners, especially Americans, in an attempt to recover forms of power perceived to have been lost in the anticolonial insurrection of Africans, Asians, Latin Americans, and indigenous peoples throughout the twentieth century, the Civil Rights Movement in the United States, the worldwide Women's Movement, and the Gay Rights Movement.

The incursion of the federal government under the George W. Bush administration into the legislative mandating of research methods marks the beginning of a new era in the politics of knowledge or the so-called science wars. Indeed, we have now moved into an era where research methodology has become a legal issue with right-wing organizations attempting to exclude scientific methods attuned to the diversity, specificity, and contextualized dimensions of human experience. In this situation, the government becomes an arbiter of what we are allowed to know. After the signing of NCLB into law in January, the Congress passed the Educational Sciences Reform Act in October 2002 to consolidate and expand the role of evidence-based research in federal education policy.

Again, as the doublespeak of the right-wing agenda reveals itself, in the name of small and unobtrusive government, the Bush administration mandates not only standardized curricula but also what methods can be used to study schools. The nation has never witnessed such restrictive forms of federal governmental control in the sphere of education. Educators are truly witnessing a science of dominant power. The right-wing educational strategy connected to this dominant power covers all its bases. At the same time that it plots to create privatized corporate schools, it makes sure that the public ones also tow the ideological line. No matter which way the struggle for privatization works out, right-wing politicos know that they will dominate schooling in America with authoritarian, antidemocratic policies that strategically eliminate anyone or any knowledge that counters their agenda (Fleischman et al., 2003; Foley and Voithofer, 2003; Lather, 2003).

The exclusive authoritarian nature of such science policies can be clearly seen in the Bush administration's *Reading First Program*. Educational leaders and researchers who raise questions about the scientific methods used to study the reading process and the performance of students and teachers in learning and teaching reading are excluded from even presenting their opinions to the Congress or the Department of Education. In the public conversation about reading and the teaching of reading that has developed around Reading First, ideological zealots have established a McCarthy-like blacklist. Long-recognized experts on reading who use qualitative research methods are no longer welcome in the community of reading scholars. At the same time, particular journals, terms, and concepts are not allowed in the conversation, as federal monies are provided only to those who pledge allegiance to the flag of positivism and the exclusive teaching of, in this case, phonics-based reading methods (Murray, 2002; Coles, 2003).

The origins of the contemporary federal educational policy that deploys a science of dominant power can be observed in right-wing educational movements of the last three decades (see Kincheloe, 1983). An immediate predecessor involves the then governor George W. Bush's educational policies in Texas in the 1990s. Numerous consultants were brought to the governor's mansion in Austin, most of whom were authors of books published by McGraw-Hill. All of the scholars and political operatives who participated in the Texas conversation called for evidence-based research that led to standardized teaching and evaluation methods in Texas schools. Since most of the research pointed to the need for McGraw-Hill textbooks, the company made a fortune in the process producing phonics-based scripted programs to be read by teachers to their classes (Trelease, 2003). In the first and second Bush administrations, this positivistic, standardized, financially lucrative process of educational research and reform has transmigrated to the federal governmental level.

Thus, while operating in the name of objective science, the right-wing educational agenda is profoundly influenced by corporate money and power. Again, positivism, privatization, and corporate influence join together in a sordid ideological ménage à trois. In fact, the Bush administration's educational proposals look like profit enhancement plans for McGraw-Hill and other corporations. In twenty-first-century education, dollars are being spent on testing, teacher manuals, and textbooks, not on efforts to promote equity and equal educational funding between rich and poor districts. The need for such a shift in funding, of course, is promoted by an evidence-based science that claims objectivity and intellectual rigor.

Indeed, such positivist research proclaims that because of the low abilities of African American, Latinos, and poor whites from these low-income districts, there is little that can be done to help them

(Kincheloe et al., 1996). Educators, the argument goes, might as well forget trying to educate such students in such a way that they can achieve socioeconomic mobility and instead focus their attention on raising the test scores of those who *are* capable of learning. Thus, NCLB mandates the creation of over 200 new tests. The federal government will spend 400 million dollars over a six-year period to develop such tests and another 7 billion dollars to implement them in all the states. The coffers of the corporate cronies runneth over.

The first day George W. Bush assumed presidency in 2001, he invited a group of so-called educational leaders to the White House. The leaders consisted mainly of Fortune 500 CEOs. A central player, of course, was Harold McGraw III, chair of McGraw-Hill. So central was McGraw to the educational reform process that he and his company had financial ties to the "objective" researchers producing the data used to justify particular Bush educational policies. A cursory reading of George W. Bush's educational policies as governor of Texas and as president always finds the McGraws at the center of decision making. Even before Bush became governor of Texas, Harold McGraw, Jr. was a board member of the Barbara Bush Foundation for Family Literacy. Bush's secretary of education, Rod Paige, was the "Harold W. McGraw Jr. Educator of the Year" during his tenure as superintendent of schools in Houston.

George W. Bush's educational relationship with the McGraws and McGraw-Hill is similar to his oil and gas relationship with Ken Lay and Enron. Indeed, the relationship between the Bushes and McGraws goes back three generations to the friendship that developed between grandfathers Sen. Prescott Bush and publishing tycoon James McGraw Jr., the uncle of Harold McGraw, Jr. The two meet in the 1930s on Jupiter Island off the east coast of Florida, an exclusive vacation spot for the northeast elite of the day. George H. W. Bush maintained the relationship with Harold McGraw, Jr. and, of course, the relation extended to the third generation with George W. Bush and Harold McGraw III. The first president Bush in the early 1990s awarded Harold McGraw, Jr. the highest award in the promotion of literacy for his profound contributions to the cause of reading. Harold McGraw III was appointed to the Bush transition advisory panel after the 2000 election. The connections between the two families go on and on as numerous Bush administration officials go back and forth between service to the president and lucrative positions at McGraw-Hill.

The influence of McGraw-Hill on the National Reading Panel's (NRP) report is a compelling example of the impact of corporate power on knowledge production about education. The NRP was commissioned by the Congress in the late 1990s to study the existing research on the teaching of reading in order to inform the contentious debate over reading pedagogy in the United States. While there are extensive problems with the report of the NRP around issues of methodology, the panel's dismissal of concerns with reading comprehension, and the panel's lack of theoretical/philosophical diversity, the most egregious problem involves the reporting of the panel's findings. The report was presented in three formats: (1) the report of the subgroups—500 pages of data including the studies on reading analyzed and the findings of the panel; (2) a 15-min video that claims to summarize the panel's findings; (3) a 32-page pamphlet that "summarizes" the larger report. Importantly, it is this pamphlet that has been the source employed by legislators to mandate reading curriculum and pedagogy.

The problem is that the short pamphlet presents recommendations for teaching reading that do not match the conclusions put forward in the report of the subgroups. The larger report warns that the teaching of phonics does not affect reading comprehension; the pamphlet in direct contradiction promotes phonics teaching maintaining that phonics instruction is the scientifically proven best method for teaching reading. It seems just a little suspicious that the NRP summary was composed in part by Widmeyer-Baker, the public relations company that McGraw-Hill employs to promote its phonics-based Open Court Reading Program. When positivism and scientific objectivity are the words of the day, such corporate influence is especially troubling.

The pamphlet, not the larger report of the NRP, has been used as the basis for educational legislation at both state and federal levels concerning the teaching of reading. In this context, the Bush administration in its first term provided 1 billion dollars a year for literacy education (the Early Literacy Initiative) for a six-year period. To administer the allocation of such monies, President Bush picked McGraw-Hill DISTAR program promoter, Christopher Doherty— DISTAR is McGraw-Hill's scripted literacy program. In light of these dynamics, McGraw-Hill has come to be known on Wall Street as a Bush stock and is showing a profound increase in its valuation because of the policies described here. Obviously, corporate-driven educational policies produce significant profits for those with political influence (*California Educator*, 2002; Karp, 2002; Yatvin 2002; Eisenhart and Towne, 2003; Metcalf, 2002; Garan, 2004).

Repackaged Positivism: Problems with Bushian Science

Positivism mixed with cronyism makes a mean cocktail. The impact of such a potent ideological potable is devastating. To make informed decisions about these politics of education, the American public simply has to better understand epistemological issues and the politics of knowledge. Here we focus in a little more detail on the specifics of Bushian educational science. A simple point grounds this analysis: connecting educational practice to educational research is a complicated and difficult task. This does not mean that it cannot be done—it most certainly can. The point is that the nature of the relationship is more complex than many—educational researchers included—realize. In teacher education, it is not easy to determine what teachers need to know about the body of research and knowledge produced in the field of education, not to mention sociology, psychology, cultural studies, history, economics, political science, literary studies, aesthetics, math, physical science, and so on. Thus, what is the relationship connecting, research, educational knowledge, and educational practice? The educational operatives in the Bush administration believe they know the answer to this question, concurrently claiming that they do not understand our warnings about its complexity.

Educational researcher Max Van Manen (1990) contends that there is a profound disconnect between educational research and educational practice. Scientific educational knowledge is in no way tantamount to pedagogical understanding. As important as I personally believe social, cultural, political, and economic contextual knowledge is to the process of teaching, I understand that simply presenting such contextual insight to teachers is not enough. There is another step that involves interpreting what such knowledge might imply for a teacher in the everyday life of her classroom. Because of the innate complexity of the knowledge and its relationship to pedagogy, I even understand that there is not one simple answer to the question concerning its implications for the classroom. However, I also know that it is a question that needs to be asked by teachers and teacher educators. The important point here is that educational research of any type does not dictate in any simple way the form educational practice should take. This, however, is the very mistake that the evidence-based research advocates of NCLB are making. Here is the research, they tell us. Now that it has been produced, teachers should go do exactly what it tells them.

Van Manen reminds us that no research "truth," no validated model of learning or "correct" teaching method can tell us what we *should* do with this student in this particular situation. Pedagogical decisions are always grounded on the specific contextual features a teacher encounters in dealing with a particular child. An emotionally sensitive child will require different approaches than one who is confident and assertive. A child who is not socially and economically privileged may require different pedagogical approaches than one who is. This list can go on and on. Positivistic evidence-based research is not interested in questions such as the ones Van Manen raises. The lived experience of students is amazingly complex and must be understood by those who seek to teach them. Van Manen (1990) puts it well:

> A child's learning experience usually is astonishingly mercurial and transitional in terms of moods, emotions, energy and feelings of relationship and selfhood. Those who absorb themselves in their children's experiences of learning to read, to write, to play music, or to participate in any kind of in or out of school activity whatsoever, are struck by the staggering variability of delight and rancor, difficulty and ease, confusion and clarity, risk and fear, abandon and stress, confidence and doubt, interest and boredom, perseverance and defeat, trust and resentment, children experience as common everyday occurrences. Parents may know and understand this reality. Some teachers do.

How often are the meanings of such microlevel, lived world experiences of children addressed by positivist researchers? The answer is simple—never. At the very least, one dimension of educational research involves exploring the intricacies of experience, the structures of meaning, the interrelationships that help construct the basic complexity of the pedagogical act.

Advocates of evidence-based research are certainly aware that sole reliance upon neopositivist research blocks any inquiry that addresses these everyday educational ambiguities. They are also aware that positivist research excludes any explorations focused on sociopolitical and justice-related concerns. There is no place for such research in the positivistic cosmos. Even prestigious groups such as the National Research Council (NRC)—an organization that has served as a scientific advisory agency to the government since 1863—discerned the need to write only its fifth report on educational research in almost a half century on the excesses of Bushian science. With the passage of educational legislation grounded on positivist ways of approaching pedagogy, the council warned of the dangerous effects of the "narrow scientism" of the Bush administration. Any rigorous educational research agenda, the NRC report asserted, should promote a wide variety of research methodologies (Lather, 2003).

The naïve view of educational data promoted by positivism assumes that the only way to determine whether knowledge produced about education is true is via replication of findings in different settings. Positivism has no mechanism to reflect or question the limitations of these replicable forms of information. This epistemology assumes the superiority of a particular type of educational knowledge that is based on finding methods that increase standardized test scores in more than one location. Of course, this makes yet another set of assumptions:

1. the knowledge included on a standardized test is neutral and objective;
2. the purpose of education involves committing such knowledge to memory;
3. educational research that does not assume the improvement of standardized test scores to be the central goal of education is not important;
4. the teaching act consists primarily of imparting this knowledge to students;
5. the measure of good teaching involves how efficiently a teacher imparts such knowledge to students;
6. being well educated is grounded on how much of this knowledge one can recall at a moment's notice;
7. the most appropriate educational format is a teacher-centered pedagogy;
8. the proper role of student is a passive knowledge receiver.

In the narrowness of what positivistic methods measure in relation to standardized test scores, issues concerning student understanding of data "learned," use of information, attitude and disposition toward learning, research skills, interpretive abilities, insight into the construction of selfhood, worldviews, and future goals, and the development of conceptual frameworks are simply irrelevant. In this context, a case certainly can be made that each of these dimensions of learning are at least as important as the "mastery" of particular content knowledge. In light of these understandings, it is important that educators, teacher educators, parents, and citizens in general make the politics of educational research an important political issue. As long as Bushian science is unchallenged, educational research will promote a right-wing political agenda all the while traveling incognito wearing the mask of objectivity (Hellstrom and Wenneberg, 2002; Coles, 2003; Eisenhart and Towne, 2003; Foley and Voithofer, 2003).

Thus, right-wing operatives have turned research design and methodology into a weapon against critically oriented teaching and learning as well as any form of pedagogy that challenges the dominant power wielders of the twenty-first century. Thus, research is transformed into an antidemocratic activity, as with the *Reading First* debacle described above. The "objective, disinterested science" of the

Bush administration simply refuses to consider analyses—such as the one presented here—of what positivism can and cannot do. We are back to the "faith-based" dimension of Bushian science—evidence-based scientific research tells us all we need to know about the educational process.

While Bush educational scientists claim all educational decisions are based on hard scientific data, they are curiously silent about the need for rigorous scientific evaluation of their own assumptions about the nature of educational science. Employing such double standards and outright duplicity, Bushian science marches forward in its application of ideological science moving from discipline to discipline in its effort to control knowledge and the curriculum grounded upon it. Yet another irony of such a politics of knowledge and educational policy is that this authoritarian process takes place under the banner of the recovery movement's call for decreased government intervention in our personal and institutional lives. Historically, there has never been as much federal control of local education as in the second Bush presidency (Lather, 2003; Street, 2003). Such authoritarianism is justified by an appeal to the authority of neutral science: this is what the best science tells what schools should be doing. How can we go against the authority of evidence-based science?

I have written about the complexity of educational science in great detail in other works (see Kincheloe, 2004a, 2004b). In these I call for a richer, more scholarly and more practical form of understanding of the types of knowledges that are needed by educators at all levels—policy makers, administrators, and teachers. The important point here is that there are many types of educational knowledge that educators need to construct: rigorous, scholarly challenging, just, contextually savvy, and humane schools. Such appreciation of the multiple knowledges is a complex epistemological concept. Educators who possess such a sophisticated understanding of knowledge recognize the following categories of educational information:

(1) Empirical knowledge comes from research based on data derived from sense data and observation of diverse dimensions of education. The positivist data of Bushian science falls under this category, but it is just one of many forms of empirical knowledge. The type of rigorous, critical education we propose insists on a far more savvy, informed, multilogical notion of empirical knowledge than the positivist ones being deployed in contemporary education. Such a thicker notion demands that researchers understand the assumptions on which their methods and designs rests, both the assets and liabilities of such methods and designs, and the multiple interpretations of the

data that are possible. Thus, a more rigorous form of empirical research understands that it does not simply, directly, and unproblematically tells us what to do in education.

(2) *Normative knowledge* concerns questions of what educators should be doing in schools, and what is the purpose of education. Such are the normative inquiries that are necessary to all pedagogical activity in a democratic, egalitarian society. Such knowledge is constructed not arbitrarily but in relation to a rigorous analysis of certain visions of a good society, questions of power and its relationship to education, and the social, cultural, historical contexts in which we operate. Normative knowledges are overtly erased in Bushian science. All of these issues are dealt with but not in the way advocated here. In positivism, such deliberations take place covertly and are tacitly embedded in the choices of research design and application—for example, a standardized test-driven curriculum contains the normative assumption that the purpose of school is to inculcate particular validated truths to all students.

(3) *Critical knowledge* is closely connected to normative knowledge as it involves the political and power-related dimensions of the educational process and how they shape what educators do. All educational decisions are political in that they are implicated in particular relationships to power. Curriculum simply cannot be constructed outside of a connection to sociopolitical power. In the knowledge we confront, in the way we deal with it, we establish particular political positions. Do we treat curricular knowledge as a final body of truth that should never be challenged? Or do we position curricular knowledge as information always open to scrutiny and rejection? Do we see the curriculum not as a body of pregiven information or one to be constructed by teachers and students in particular locations? All of these curricular approaches assume a particular relationship with power whether we want them to or not. Discerning the nature of these power relations is a central concern of critical educational knowledge producers.

(4) *Ontological knowledge* has to do with what it means to *be* a teacher. Ontology is the branch of philosophy that studies what it means to be in the world, what it means to be human. Thus, ontological educational knowledge is concerned with the way teachers come to see themselves as professional educators, how they develop their teacher personas. In a critical education, such knowledge is profoundly important because teachers with differing teacher personas will hold diverse views of educational purpose, different interpretations of educational knowledges, different expectations of and relationships with students, different reasons for becoming a teacher and

staying in the profession. Such ontological knowledge is produced as researchers and teachers themselves explore their own and education's relationship to the social, cognitive, cultural, philosophical, political, economic, and historical world around them.

(5) *Experiential knowledge* is often recognized as the most important form of knowledge in craft-based views of the teaching profession. The mistake often made in such a framework is that experiential, hands-on knowledge of the teaching act becomes the only knowledge valued. A critical, democratic education profoundly values the importance of experiential knowledge, but understands that it must be accompanied by empirical, normative, critical, ontological, and reflective synthetic educational knowledges. Experiential knowledges are inherently complex because there is so much disagreement as to what constitutes practice. In the spirit of our multilogicality, there are many types of practices and thus many types of experiential knowledges that emerge from them. Knowledge derived from practice, while necessary, is never self-evident. Because of what Donald Schön describes as "indeterminate zones of practice," experiential knowledge—like empirical knowledge—does not generalize well. Since professional practice is always marked by surprises that force practitioners to reshape their understandings of particular situations and ways of operating, teachers learn to use their experiential knowledge to devise not rigid rules but improvisational orientations toward their professional activity.

(6) *Reflective-Synthetic knowledge* is the knowledge base of education that includes all of the forms mentioned here—and more. Professional work in educational science demands reflection on their multiple relations to practice, broadly defined, and a synthetic process that studies the implications of the different forms of knowledge when juxtaposed to one another. Here rests a key dimension of the type of complex educational science that frees us from the failure of positivism and the recovery of positivism in the evidence-based cosmos described in this chapter. A complex educational science produces diverse forms of knowledge and then takes on the difficult job of analyzing them in relation to one another, discerning all the while their connection to educational practice. Devising more rigorous, more beneficial, more pragmatic ways of accomplishing this task is a central goal of my own work and the authors in this collection of essays.

In light of the diverse forms of educational knowledges and the need for educators to synthesize and apply them, the claims of the Bush administration that invalidated teaching practices and "unproven education theories" are the main reasons for school and student failure are seen more clearly as ideologically driven doublespeak. Such claims are a form of scientific mumbo jumbo that use a scientific language to claim validity for particular political positions. On what basis are particular teaching practices invalid? Using what criteria? Employing what epistemology? What normative assumptions? What is an unproven educational theory? Is an educational policy that bases educational success in part on all students' success in engaging in sophisticated knowledge work and demonstrating the ability to conduct primary research an unproven educational theory?

Such a normative form of knowledge is not empirically provable or disprovable. When I argue that education should be overtly antiracist how do we empirically prove that such an assertion is empirically true or false? The answer—we cannot. An epistemologically informed education scholar can make such knowledge-based distinctions. Such a scholar knows that different educational questions demand different forms of educational knowledge. In such a context this informed scholar fights against the effort to privilege positivist experimental research, evidence-based research, and randomized trials as superior to all other forms of inquiry (Eisenhart and Towne, 2003; Foley and Voithofer, 2003). Right-wing operatives bank on the belief that since these questions are so complex and the use of the term, scientific, holds such political capital, few people will study the details of the issues in question. Such a position is not only cynical but also reveals a political orientation that is inherently uncomfortable with democracy and democratic institutions.

Conclusion: The Future of Democratic Education

The authoritarian educational and epistemological fundamentalism aided and abetted by corporate cronyism in the Bush administration is merely one dimension of a larger radical right-wing agenda including market-driven, sociopolitical norms, redistribution of wealth from poor to rich, globalized economic imperialism, religious zealotry, and militaristic empire building around the globe. The control of education and research is a central dimension of this larger agenda, as it helps to repress the citizenry's access to diverse perspectives about these matters. In this context, such educational policies can subvert the civic dimensions of schooling—the production of critical, analytical citizens who have knowledge of a wide variety of perspectives on diverse topics and who are dedicated to participation in democratic practices. These right-wing policies cannot be implemented without an uninformed population, who can be frightened into supporting a political and educational status quo that works against their interests.

Indeed, Chester Finn, the president of the right-wing Fordham Foundation, has produced research that "proves" that the American public does not want schools to graduate students who engage in these acts of democratic citizenship. Americans, Finn argues, want schools that teach only what he calls "fundamental knowledge," meaning that which always supports dominant power as it works to preclude dissent. With bills such as House Resolution 3077 passing in the fall of 2003 mandating that the government monitor international relations classrooms to detect those that are not operating in the "national interest," the effort to control knowledge in educational institutions becomes even more outlandish (Doumani, 2004; WEAC, 2004). Of course, a critical politics of knowledge understands that no knowledge—obviously, present knowledge included—is disinterested. In this epistemological context, however, I am much more comfortable with knowledge producers who recognize this dimension of inquiry than those who believe that what they produce is the "objective truth."

Education scholars who understand this dynamic are far more likely to appreciate the need for multiple forms of knowledge in policy-making contexts. To stack the deck with any monolithic theoretical perspective is misguided in public policy making. A central dimension of a critical mode of public policy making involves the interaction of diverse perspectives in the marketplace of ideas. This is exactly what is *not* happening in the authoritarian knowledge production operations of contemporary right-wing educational politics. What Americans do not know is that the public availability of alternate modes of knowledge production and outlets for diverse opinions are fading in the long shadows of corporate controlled, privatized information. The politics of knowledge must become a primary educational and political issue in the coming years if democracy, not to mention a civic-minded education, is to survive.

References

Apple, M. (1993). The Politics of Official Knowledge: Does a National Curriculum Make Sense? *Teachers College Record*, 95, 2, pp. 222–241.

Aratani, L. (2004). States Criticize Bush Education Plan. http://www.mercurynews.com

Berkowitz, B. (2001). Public Schools Open for Business. http://www.workingforchange.com/article.cfm

Bogle, C. (2003). Cuts in Education Funding Will Improve Academic Performance. Honest. http://www.wsws.org/articles/2003/aug2003/educ-a28.shtml

California Educator (2002). Scripted Learning: A Slap in the Face? 6, 7. http://ww.cta.org/californiaeducator/v6:7feature_4.htm

Coles, G. (2003). Learning to Read and the "W Principle." *Rethinking Schools*, 17, 4. http://www.rethinkingschools.org/archive/17_04/wpri174.shtml

Coontz, S. (1992). *The Way We Never Were: American Families and the Nostalgia Trap.* New York: Basic Books.

Doumani, B. (2004). *Personal Voices: The End of Academic Freedom.* AlterNet. http://www.alternet.org/story18426

Eisenhart, M. and L. Towne (2003). Contestation and Change in National Policy on "Scientifically Based" Education Research. *Educational Researcher*, 32, 7, pp. 31–38.

Fleischman, S., J. Kohlmoos, and A. Rotherham (2003). From Research to Practice: Moving Beyond the Buzzwords. http://www.nekia.org/pdf/ed_week_commentary.pdf

Foley, A. and R. Voithofer (2003). Bridging the Gap? Reading the No Child Left Behind Act against Educational Technology Discourses. http://www.coe.ohio-state.edu/rvoithofer/papers/nclb.pdf

Garan, E. (2004). *In Defense of Our Children: When Politics, Profit, and Education Collide.* Portsmouth, NH: Heinemann.

Gonzalez, R. (ed.) (2004). *Anthropologists in the Public Sphere.* Austin, TX: University of Texas Press.

Gresson, A. (1995). *The Recovery of Race in America.* Minneapolis, MN: University of Minnesota Press.

Gresson, A. (2004). *America's Atonement: Racial Pain, Recovery Rhetoric, and the Pedagogy of Healing.* New York: Peter Lang.

Hartman, A. (2002). Envisioning Schools beyond Liberal and Market Ideologies. *Z Magazine*, 15, 7. http://www.zmag.org/amag/articles/julang02hartman.html

Hellstrom, T. and S. Wenneberg (2002). The "Discipline" of Post-Academic Science: Reconstructing the Paradigmatic Foundations of a Virtual Research Institute. http://www.cbs.dk/departments

Hursh, D. (2001). Standards and the Curriculum: The Commodification of Knowledge and the End of Imagination. In J. Kincheloe and D. Weil (eds.), *Standards and Schooling in the United States: An Encyclopedia*, pp. 735–744. Santa Barbara, CA: ABC-Clio.

Karp, S. (2002). Let Them Eat Tests. *Rethinking Schools*. http://www.rethinkingschools.org/special_reports/bushplan/eat164.shtml

Kincheloe, J. (1983). *Understanding the New Right and Its Impact on Education.* Bloomington, IN: Phi Delta Kappa.

Kincheloe, J. (2004a). The Knowledges of Teacher Education: Developing a Critical Complex Epistemology. *Teacher Education Quarterly*, 31, 1, pp. 49–66.

Kincheloe, J. (2004b). The Bizarre, Complex, and Misunderstood World of Teacher Education. In J. Kincheloe, A. Bursztyn, and S. Steinberg (eds.), *Teaching Teachers: Building a Quality School of Urban Education.* pp. 1–50. New York: Peter Lang.

Kincheloe, J. and S. Steinberg (eds.) (2004). *The Miseducation of the West: How Schools and the Media Distort Our Understanding of the Islamic World*. Westport, CT: Praeger.

Kincheloe, J., S. Steinberg, and A. Gresson (eds.) (1996). *Measured Lies: The Bell Curve Examined*. New York: St. Martin's Press.

Kitts, L. (2004). Keep Special Interests out of America's Classrooms. http://www.hoodrivernews.com/lifestyle%20stories/067%20special%20interest%20opinion.htm

Lather, P. (2003). This IS Your Father's Paradigm: Government Intrusion and the Case of Qualitative Research in Education. http://www.coe.ohio-state.edu/plather/

Metcalf, S. (2002). Reading between the Lines. *The Nation*. http://www.lindaho.yt.com/title%20I.htm

Miner, B. (2004). Why the Right Hates Public Education. *The Progressive*. http://www.progressive.org/jan04/miner0104.html

Murray, A. E. (2002). Reading's New Rules: ESEA Demands a Scientific Approach. *Education Update*. http://www.ascd.org/publication/ed_update/200208/murray.html

Ohanian, S. (1999). *One Size Fits Few: The Folly of Educational Standards*. Portsmouth, NH: Heinemann.

Steinberg, S. and J. Kincheloe (eds.) (2004). *Kinderculture: The Corporate Construction of Childhood*. 2nd edn. Boulder, CO: Westview.

Street, B. (2003). What's "New" in New Literacy Studies? Critical Approaches to Literacy in Theory and Practice. *Current Issues in Comparative Education*, 5, 2.

Trelease, J. (2003). All in the Family. http://www.trelease-on-reading.com/whatsnu-bush-mcgraw.html

Van Manen, M. (1990). *Researching Lived Experience: Human Science for an Action Sensitive Pedagogy*. Albany, NY: State University of New York Press.

Vinson, K. and E. Ross (2001). Social Studies—Social Education and Standards-Based Reform: A Critique. In J. Kincheloe and D. Weil (eds.), *Standards and Schooling in the United States: An Encyclopedia*, pp. 909–928. Santa Barbara, CA: ABC-Clio.

Weil, D. (2001). Functionalism—From Functionalism to Neofunctionalism and Neoliberalism: Developing a Dialectical Understanding of the Standards Debate through Historical Awarenss. In J. Kincheloe and D. Weil (eds.), *Standards and Schooling in the U.S.: An Encyclopedia*. Santa Barbara, CA: ABC-Clio.

Wisconsin Education Association Council (WEAC) (2004). The American Board and Fast Track Certification: An Attack on the Teaching Profession. http://ww.weac.org/pdfs/2003-2004/certification_research.pdf

Yatvin, J. (2002). Babes in the Woods: The Wandering of the National Reading Panel. *Phi Delta Kappa*, 8, 5, pp. 364–369.

Chapter 2

How Did this Happen? The Right-Wing Politics of Knowledge and Education

Joe L. Kincheloe

In chapter 1, I explored the contemporary right-wing threat to democracy and to democratic education. Over the last few decades, right-wing operatives have constructed a new view of the world where right-wing principles, America, the basic tenets of Western civilization, fundamentalist Christianity, positivist science, white supremacy, and patriarchy are deemed to be under attack by foreign and domestic enemies. In this configuration, the American Christian heterosexual white male is the primary victim of the new world order. This construction of the victimization of the American Christian white male plays a central role in American electoral politics in the first decade of the twenty-first century. In the "red states" of the 2000 and 2004 elections, Christian white males voted overwhelmingly in favor of George W. Bush. The Bush campaigns in these two elections plucked the heartstrings of this constituency, as they promised protection from the assaults of affirmative action, multicultural curricula, anti-American professors, gay marriage, and anti-Christian values. In many ways the right-wing/fundamentalist Christian rise to power in America over the last 30 years can be viewed as an effort to defend the faith against the attacks against the West (as embodied by America) and its God.

The defensive consciousness produced by such a perspective catalyzes numerous social, cultural, educational, and even geopolitical policies and actions. In an educational context, such a consciousness shapes the nature of the classrooms we can envision and bring into existence. The right-wing worldview driving contemporary school policy in the first decade of the twenty-first century is characterized by

1. positivism;
2. indoctrination of Western/American superiority;

3. belief that intelligence is genitically determined;
4. universally valid knowledge; one "truth" that is universally valid;
5. standardized curricula;
6. an emphasis on low-level cognitive activities; rote memorization;
7. the reality of an isolated individual removed from social, historical, and cultural context;
8. a fear of multiple points of view and dissent.

Developing a Historical Consciousness

A central task of this chapter (and the book) is to help us understand in a historical context why so many Americans have bought into these regressive, oppressive, and anti-democratic politics and educational activities. One of the failures of the contemporary political and educational conversation is that we don't examine pressing issues in larger contexts. Acceptance of the retrograde policies described in Chapter 1 can only be explained in light of one the most dominant sociopolitical and philosophical dynamics of the last 500 years—European, and especially in the last 100 years, American colonialism.

Though it is rarely discussed in relation to education, the sociopolitical, philosophical, psychological, and economic structures constructed by the last 500 years of Euro-American colonialism have a dramatic, everyday affect on what goes on in classrooms. After several centuries of exploitation, the early twentieth century began to witness a growing impatience of colonized peoples with their sociopolitical, economic, and educational status. A half millennium of colonial violence had convinced Africans, Asians, Latin Americans, and indigenous peoples around the world that enough was enough. Picking up steam after World War II, colonized peoples around the world threw off colonial governmental strictures and set out on a troubled journey toward independence. The European colonial powers, however, were not about to give up such lucrative socioeconomic relationships so easily. With the United States leading the way, Western societies developed a wide-array of neocolonial strategies for maintaining the benefits of colonialism. This neo-colonial effort continues unabated and in many ways with a new intensity in an era of transnational corporations and the "war on terror" in the twenty-first century.

Understanding these historical power dynamics and their influence is central to our metahistorical consciousness. Indeed, though most Americans are not aware of it, the anticolonial rebellion initiated the liberation movements of the 1960s and 1970s that shook the United

States and other Western societies. The Civil Rights Movement, the women's movement, and the gay rights movement all took their cue from the anticolonial struggles of individuals around the world. For example, Martin Luther King wrote his dissertation on the anticolonial rebellion against the British led by Mohandas Gandhi in India. King focused his scholarly attention on Gandhi's nonviolent colonial resistance tactics, later drawing upon such strategies in the civil rights movement.

By the mid-1970s, a conservative counterreaction—especially in the United States—to these liberation movements was taking shape with the goals of "recovering" what was perceived to be lost in these movements (Gresson, 1995, 2004; Kincheloe et al., 1998; Rodriguez and Villaverde, 2000). Thus, the politics, cultural wars, and educational and psychological debates, policies, and practices of the last three decades cannot be understood outside of these efforts to "recover" white supremacy, patriarchy, class privilege, heterosexual "normality," Christian dominance, and the European intellectual canon. They are some of the most important defining macro-concerns of our time, as every social and educational issue is refracted through their lenses. Any view of education conceived outside of this framework becomes a form of ideological mystification. This process of ideological mystification operates to maintain present dominant–subordinate power relations by promoting particular forms of meaning making. In this colonial context, ideological mystification often involves making meanings that assert that non-European peoples are incapable of running their own political and economic affairs and that colonial activity *was* a way of taking care of these incapable peoples.

Contemporary standardized, test-driven, psychologized education is enjoying great support in the twenty-first century because, in part, it plays such an important role in recovering what was perceived to have been lost in the anticolonial liberation movements. One of the educational dimensions of what was perceived to have been lost involves the notion of Western or white intellectual supremacy. No mechanism works better than intelligence/achievement testing and school performance statistics to "prove" Western supremacy over the peoples of the world. Positivistic psychometricians operating in their ethnocentric domains routinely proclaim the intellectual superiority of Western white people. Statistically, white students perform better in schools than nonwhite students. Richard Herrnstein and Charles Murray (1994), for example, write unabashedly that the average IQ of African people is about 75.

The fact that the concept of an intelligence test is a Western construct with embedded Western ways of understanding the world is never mentioned in this brash assertion. Thus, the contemporary educational obsession with labeling, measuring, and victim blaming is concurrently a macro-historical, meso-institutional, and a micro-individual matter. In this context, we can begin to understand the way that this labeling and measuring works to justify the colonial and neocolonial process. Marginalized people do not do well in schools not because of the social effects of their marginalization but because they are inferior. We (American upper-middle/upper class white people) are the superior beings who must "take care" of the rest of the world. Getting oil concessions and other cheap natural resources necessary to run our economy in the process is merely a well-deserved reward for our selfless efforts to help those in need.

The Recovery Movement and Its Educational Consequences

By the 1970s, right-wing educational policy was directly connected to the larger recovery movement, as it sought to eliminate the anticolonial, antiracist, antipatriarchal, and diversity affirming dimensions of progressive curriculum development. Understanding the way some educators were using education to extend the goals of the worldwide anticolonial movement and the American liberation movements in particular, right-wing strategists sought to subvert the public and civic dimensions of schooling. Instead of helping to prepare society for a socially mobile and egalitarian democracy, education in the formulation of the right-wing recovery redefined schooling as a private concern. The goal of this private concern was not to graduate "good citizens" but to provide abstract individuals the tools for socioeconomic mobility.

The progressive idea of helping marginalized *groups*, such as African Americans become socially mobile, was not the same goal as facilitating individual mobility. In fact, the two attempts often came into direct conflict. In the right-wing recovery project, the promotion of the mobility of marginalized groups was a form of social engineering that perverted the basic goals of education. The promotion of the mobility of abstract individuals in this conceptual context was a tribute to the basic American value of meritocracy. Only the intelligent and virtuous deserved mobility and such individuals according to the recovery movement's cognitive theorists, Richard Herrnstein and Charles Murray (1994) of *bell curve* fame, tended to be white and upper-middle class. Employing the rhetoric of loss, the promoters of recovery spoke of the loss of standards, discipline, civility, and proper English. Because of the pursuit of racial/cultural difference and diversity, America itself was in decline. In the rhetoric of recovery, the notion of loss and falling standards was always accompanied by strategically placed critiques of affirmative action, racial preferences, and multiculturalism. Though the connection was obvious, plausible deniability was maintained—"we are not racists, we only want to protect our country from the destruction of its most treasured values."

By the 1970s, with the emergence of this ideology of recovery, the very concept of government with its "public" denotations began to represent the victory of minorities and concerns the inequities of race, class, gender, and colonialism. "Big government" began to become a code phrase for antiwhite male social action in the recovery discourse. Indeed, in this articulation, it was time to get it off "our" backs. Thus, privatization became more than a strategy for organizing social institutions. Privatization was the ostensibly deracialized term that could be deployed to signify the recovery of white, patriarchal supremacy. In the same way, the word, choice, could be used to connote the right to "opt out" of government mandated "liberal" policies. Like good consumers, "we" (Americans with traditional values) choose life, privatized schools, the most qualified job applicants, and Christian values over other "products."

Thus, in the grander sense, we choose the private space over the *diversity* of the public space. In rejecting the public space, the right wing rejected the political domain—a choice that resonated with many conservative white Christians throughout the nation. Indeed, any political action on our part, the advocates of recovery asserted, will in effect be antipolitical. We will work to make sure that traditional "political types" be defeated by antigovernment agents who will work to undermine the public space with its social programs, infrastructures, and, of course, schools. Thus, we witness a decline in interest in the political and the academic. Indeed, politicians who are not born-again Christians working to dismantle the public space and academics who are not denouncing the academy are not our type of people. In the recovery, the institutions of public government and education must go. Both institutions, the right-wing argument goes, display the tendency to undermine the best interests of fundamentalist white people—white males in particular.

Thus, in this historical context, we can better understand the right-wing use of No Child Left Behind (NCLB) federal law as a legal tool to reconfigure the federal government's role as the promoter of equality and diversity in the educational domain. Though it was promoted as a new way of helping economically marginalized and minority students, such representations were smokescreens used to conceal its mission of recovery of traditional forms of dominant power. In this power context, NCLB is quite cavalier about the inequity between poor and well-to-do school districts and even schools within particular districts. The right-wing public discourse about education has successfully erased questions of race and class injustice from consideration. The fact that 40 percent of children in the United States live in poor or low-income conditions is simply not a part of an educational conversation shaped by the rhetoric of recovery. The understanding that students who are upper-middle class and live in well-funded schools and/or school districts have much more opportunity for academic and socioeconomic success than students from poor contexts is fading from the public consciousness in the twenty-first century.

The realization that inequality is deemed irrelevant even when we understand that socioeconomic factors are the most important predictor of how students perform on high stakes standardized tests, is distressing. In this context, we begin to discern that in a system driven by such high stakes tests, it is not hard to predict who is most likely to succeed and fail. In the name of high standards and accountability, the recovery project scores great victories. "We can't let these 'incompetents' get by with such bad performance," right-wing ideologues righteously proclaim, "it degrades the whole system." As they cry their crocodile tears for poor and marginalized students in their attempt to hide their real agenda and garner support of naïve liberals for their educational plans, they concurrently support deep cuts in any program designed to help such students.

During the George W. Bush presidency, for example, Americans have witnessed cuts in food stamps; Temporary Assistance for Needy Families; nutrition programs for children; childcare; the enforcement of laws for child support, child health insurance, childcare; and the Low-Income Home Energy Assistance Program. And this does not include the education programs that help poor and marginalized students targeted for termination in the coming years. The privatization-based voucher programs proposed as a means of helping students from poor families avoid failing schools and gain access to a higher quality education do not work. The price of attending many private schools, especially the elite ones, is more costly than the worth of the meager voucher. Most students from poor families even with their vouchers will still not be able to afford private education, not to mention meeting the high standardized test score requirements such schools require. Such issues are, of course, not typically a part of the truncated public conversation about vouchers and private schooling.

None of the right-wing educational proposals deal honestly with issues of inequality. With a wink and a nod, they offer suggestions that have little to do with the profound labor needed to help improve the possibility of academic success for poor and racially marginalized students. The Heritage Foundation, for example, responding to the question, how do we improve marginalized student school performance and help get them out of poverty, suggests to

1. get rid of "progressive education" and in its stead demand basic skill teaching—progressive education is defined here as any pedagogy that starts "where students are" taking into account student needs rather than imposing a standardized curriculum from outside;
2. promote high stakes testing;
3. replace principals who complain about not having enough funds with ones who do not;
4. fire staff who do not believe in the mission of such traditional forms of schooling.

Such suggestions serve the recovery ideology well, as they guarantee the underfunding of poor schools, the use of failed pedagogies, and the failure of marginalized students. With such policies in place, we can scientifically "certify" the inferiority of students from disenfranchised backgrounds. The "naturally superior" will take their proper places in the scientific, technological, academic, and professional marketplace. Meritocracy will have worked, right-wing ideologues will proclaim (Hartman, 2002; Karp, 2002; Coles, 2003).

Such faux-meritocratic educational policies are designed to "fix" the academic race. Standardized curricula and standards-based assessments not only censor diverse perspectives (read, critical), but they also make sure the culturally and socioeconomically privileged have their privilege officially validated. Indeed, several researchers have identified a tendency for poor and minority students to drop out at higher rates as standardized test scores rise (McNeil, 2000; Horn and Kincheloe, 2001). In the name of standards and quality education, minor and easily addressed intellectual characteristics of students of color take on monumental importance. Verb-ending usage by some African American and Latino students becomes "empirical proof" of their writing problems and even English language deficiency (Fox, 1999). No matter how brilliant other dimensions of their writing and language usage may be, they are often described as not being "academic material."

I have known of or have taught scores of minority students who brought such writing tendencies to school with them but quickly dealt with them when given a chance. Understanding such tendencies in larger socioeconomic and cultural context, they came to appreciate how such cultural characteristics would be unfairly used against them and other African American, Latino, and Native American students. In recovery grounded educational contexts, existing forms of inequality are allowed to continue and with the implementation of NCLB and standardization policies, new forms of inequity are developing. Educators concerned with promoting rigorous academic work along with understanding and help for economically and culturally marginalized students face institutionalized obstacles in the remaining years of the first decade of the twenty-first century. The recovery of white supremacy, patriarchy, and class elitism has entered a new educational phase in the era of NCLB and other George W. Bush educational policies.

The Recovery of the Supremacy of Western Knowledge

A central dimension of the right-wing recovery movement in the United States involves the revalidation of Western ways of seeing the world and producing knowledge. Such modes of scholarship were criticized during the anticolonial movements and the academic expression of such resistance in postcolonial and post-structuralist forms of inquiry. Postcolonialism and post-structuralism consistently challenged the universality of Western knowledge production and its capacity to oppress those individuals who failed to fit the certified criteria that emerged from such colonial scholarship. Those children, for example, from African villages who did not fit Western psychology's universal stages of cognitive development and were deemed "slow" or "developmentally challenged." Such children were not deficient—just different from the culture that produced the *universal* stages. Of course, those who were culturally, racially, or linguistically different in the United States have often faced these same dynamics. From my perspective it is not hard to understand why many scholars and educators rebelled against this oppressive regime of truth.

The central feature of the colonial knowledge we watch being recertified in the recovery movement involves the superiority of the West. Right-wing operatives in the recovery of Western supremacy have to wipe out all of the knowledge and memory that would undermine this West-is-best sentiment. Thus, attempts to include the study of Western colonial interactions with the Islamic world after 9/11, for example, have to be proclaimed "anti-American." Efforts to study power and the way it operates in the contemporary corporatized mediascape have to be subverted by right-wing interest groups. In a standardized or corporatized educational system, of course, we need not worry about such knowledges and teachings reaching our young. In a corporatized media they will not reach our old people either. Once we know these things about education, we can begin to point out the values, the assumptions, the privileged and excluded voices, the historical inscriptions found within all knowledge. Such a task allows us to better understand the insights and limitations of Western rationality and how they can work to both empower and disempower those they encounter. These are the activities that constitute rigorous scholarship. In the types of education being championed by the right wing, they are erased. The Western worldview is beyond questioning in the recovery movement. In addition to promoting particular perspectives on the nature of the physical world and society, worldviews dictate what phenomena can be known as well as the process by which they can be known.

Those of us who want to study the workings of power, for example, via the deployment of signifiers and appeals to the unconscious know that such subtle processes fall outside the parameters of dominant ways of producing knowledge. As such complex modes of understanding are excluded from legitimate scientific inquiry, dominant forms of power continue to work at a level invisible to most people. In this way, dominant positivistic ways of seeing contributes to oppression of those who are "different" while allowing dominant science to continue its journey down a colonial path. Such a trek illustrates an intellectual and moral stagnation that is camouflaged by the recovery movement (Grossberg, 1992; Keith and Keith, 1993; Hess, 1995; Woodhouse, 1996).

It is in this stagnant context that I have proposed the notion of bricolage (Kincheloe, 2001; Kincheloe and Berry, 2004). The French word *bricoleur* describes a handyman or handywoman who makes use of the tools available to complete a task. In the context of scientific research, I use this term to denote the process of employing multiple research methodological strategies and theoretical discourses as they are needed in the unfolding context of the research situation. The point is to get beyond the monological colonialistic perspective of positivism and engage new ways of understanding from diverse intellectual and cultural traditions. The bricoleur works diligently to uncover the hidden artifacts of power and the way that shapes the knowledge that researchers produce. Of course the diversity, of the bricolage offers an alternative to the monocultural ways of perceiving protected by the recovery movement.

Without the multiple perspectives of the bricolage or something akin, it is doubtful whether the stagnant methods of positivism will move us to new insights that appreciate the multiple diversities of the planet. Without such forms of critique and action, the recovery movement will operate to create institutions and modes of consciousness that will protect the privilege of dominant groups while certifying the "deficiencies" of marginalized peoples. Multiple perspectives on knowledge as well as multiple sources of knowledge are needed to overcome this cultural and political dominance of Western ways of seeing. This is why it is so important in right-wing educational policy to quash these types of epistemologies and curricula. In addition, this is why such concepts as local knowledges, subjugated knowledges, and indigenous knowledges—information produced outside the boundaries of positivism and Western ways of seeing—are so threatening to dominant power. From a critical multilogical position, valuing such ways of knowing and the knowledges they produce is akin to valuing biodiversity—awareness of this epistodiversity grants us new insights into the world and our role in it.

Without this epistodiversity, we are tied to an "evidence-based" positivist form of knowledge production riddled with harmful assumptions that often undermine the possibility of sustainable human life in sustainable environments socially grounded on democratic and egalitarian principles. As a colonial epistemology, positivism has traditionally produced knowledges needed by administrators and managers of political institutions, corporations, and the military. In this context, positivistic knowledges have been deployed for the purpose of maintaining the empire: economic hegemony, political domination, and patriarchal oppression (Harding, 1996). Understandings that challenge these power relations are difficult to produce via the epistemologies and conceptual frameworks constructed around imperial administrative and managerial tasks.

Colonial Knowledge: Right-Wing Education as Recovery of Dominant Power

In such positivist frameworks, the natural world has been constructed as a passive and inert entity that needed to be classified and ordered for the purpose of domination. The right-wing educational reforms of the contemporary era "recover" the domination impulse and reinsert it into the sphere of teaching and learning. Indeed, in this context, all teachers and learners must be classified as either effective/intelligent or incompetent/slow. To preclude the possibility of teacher incompetence, all teaching must be ordered—that is, standardized and controlled. In this positivist framework, new forms of inequity are produced, as educational research about inequality is brushed aside as are forms of teaching and curriculum development that work to promote educational justice. The right-wing recovery creates an intellectual climate where America has become uninterested in questioning itself. This allows for the growth of a conservative absolutism that promotes the West-is-best—particularly the U.S.-is-best—mindset devoted to free market economics, globalized economic imperialism, geopolitical expansionism, and education as a celebration of U.S. supremacy and moral superiority (Bogle, 2003; Foley and Voithofer, 2003; Kitts, 2004).

The idea that U.S. economic, geopolitical, and educational policies are all interrelated and mutually supportive is something that many Americans do not know. The goal of educating critical democratic citizens who ask hard questions about the ethical dimensions of both America's role in the world and its global and domestic economic policies simply does not fit the mission of the recovery. In fact, the work of democratic citizens in general may not fit such a mission. In the rhetorical universe of the recovery movement, asking hard questions of American actions is deemed an "anti-American activity." The recovery movement's politics of knowledge are vicious and deadly serious about subverting critique of contemporary U.S. actions at home and in the world.

In spring 2002, for example, leaders of the Bush Department of Education issued orders to delete material from the 30-year-old Educational Resources Information Center (ERIC) database that does not support the general philosophy of NCLB. Every assistant secretary of education was directed to form a group of departmental employees with a least one person who "understands the policy and priorities of the administration" to scrub the ERIC web site. Such action runs counter to the original intent of the web site established in 1993 to construct a permanent record of educational research for students, teachers, citizens, educational researchers, and other scholars. Concurrently, such information deletion raises the stakes of right-wing knowledge politics to a new level, as individuals will only have access to public data that supports particular ideological agendas. Such actions are unacceptable in a democratic society (The Memory Hole, 2002; OMB Watch, 2002; Lather, 2003).

In place of the "discredited" research found on the ERIC web site and many other locales, the new Department of Education's Institute of Education Sciences in August 2002 created a web-based What Works Clearinghouse project. The project is promoted as a one-stop source of evidence-based teaching methods required by NCLB. Here educators will gain access to exclusively positivistic data in an ideological effort to shape the conversation about education as well as educational practice itself (Street, 2003). One will not find analyses of the politics of knowledge or the relationship between larger geopolitical policies and Bush administration's educational agenda here. Indeed, one will be hard pressed to find anything about the social, cultural, or political context of education. Such analysis does not fall under the category of scientific research about education.

Thus, returning to our discussion of positivism in chapter 1, the Bush administration under the cover of the objectivity of scientific research shapes and controls knowledge, keeps teachers disempowered, and positions education as a source of right-wing indoctrination for years to come (Hartman, 2002; Kitts, 2004). If such policies succeed, no one will use education as a means of assessing the status quo or of questioning the geopolitical and domestic paths on which America presently finds itself. Thus, the standardization and privatization efforts referenced in chapter 1 are deployed as key weapons in the right-wing politics of knowledge. Such efforts can help guarantee a monolithic curriculum and subvert all forms of multilogicality and epistodiversity from the educational process. Make no mistake, teachers in this ideological configuration are distributors of prepackaged information—not producers or interpreters of knowledge. They are functionaries who are told how and what to teach—canon fodder in the grand recovery movement.

Why Is the Rest of the World Reacting so Negatively to the Recovery Agenda?

The right-wing politics of knowledge operating in the United States in the twenty-first century makes for a world where reality is perceived across a great chasm of misconstruction (Sarder, 1999). Distorted pictures of an irrational and barbaric people shape decision making in areas of foreign policy, economics, and education. As discussed above positivistic ways of seeing in the educational, geopolitical, economic, and cultural spheres while claiming neutrality are profoundly shaped by discursive, ideological, and historical contexts. In the Second Gulf War and the public debate surrounding it, reporters for the major U.S. television networks denied that their coverage was framed by particular neo-conservative perspectives. American television was objective and fair while Qatar's Al-Jazeera was biased and characterized by low journalistic standards.

One of the lessons that scholars around the world learned in the last third of the twentieth century was that no knowledge is disinterested. All information is produced by individuals operating at a particular place and a specific time—they see the world and employ methods for viewing the world from a particular point in the complex web of reality.

Yet, the right-wing politics of knowledge examined here refuses to consider this epistemological concept. Individuals from around the world are often shocked by the American right wing's refusal to examine their own knowledge production, their perspectives on America, and their view of the rest of the world. They are especially shocked by the right-wing ignorance of the U.S. role in the world. When right wing educator Chester Finn (2002) writes that 9/11 presented a chance for Americans "to teach our daughters and sons about heroes and villains, about freedom and repression, about hatred and nobility, democracy and theocracy, about civic virtue and vice," the American blindness to colonial and neo-colonial atrocities around the world is revealed.

Indeed, Finn, George W. Bush, Dick and Lynne Cheney, and other purveyors of the right-wing politics of knowledge simply refused to recognize that 9/11 in part reflected the rage toward the United States pulsing through the veins of many Muslims. The indifference displayed by many U.S. policy makers toward the suffering of everyday people around the Islamic world fanned the flames of this anti-American fury. In Iraq, for example, the indifference of American leaders to the effects of the post–First GulfWar sanctions put into place in 1991 angered millions of Muslims around the world as well as the Iraqi people (Sudetic, 2002). This is one of many reasons that when U.S. and British forces invaded the country in March 2003, they were not met with flowers and kisses of a grateful people that George W. Bush promised Americans. Most Iraqis obviously did not see the Second GulfWar as the War of Iraqi Liberation despite their disdain for Saddam Hussein. The fires of the Iraqi insurgency were fired by this anger toward American ways of seeing them, and the subsequent actions such perspectives promoted.

The fundamentalism of the right-wing politics of knowledge is central to our understanding of the growing hatred and mistrust of the United States Fundamentalism as used in this context is defined as the belief in the ultimate superiority of Americana, the American political philosophy in particular as well as the Western scientific creed and its methods for producing objective knowledge—positivism writ large. The Fordham report well illustrates this fundamentalism albeit in a manner that avoids unambiguous statement of its position. The ideology and rhetoric of the report make it a document well worth the analysis. Produced by Chester Finn's right-wing Fordham Foundation, the Fordham report was issued on the one-year anniversary of 9/11. Entitled, "September 11: What Our Children Need to Know," assumes from the beginning that Americana and American political philosophy are monolithic expressions of one culture's social and political values. The United States is not now nor has it ever been monocultural and of one political mind. The effort to construct the assumption that America has a common culture and politics is an attempt to position particular ethnicities and specific political points of view as existing outside the boundaries of true Americanism. When educators focus on diversity and multilogicality, the right-wing logic posits, they are misleading their students and pushing a relativistic agenda where nothing is right or wrong. The epistemological naïveté of such assertions is blatant, as Fordham authors ignore an entire body of social theoretical work that moves far beyond the polar extremes of pure objectivity on one end of the continuum and relativism on the other (Gadamer, 1975; Madison, 1988; Van Manen, 1991; Kincheloe, 2001; Thayer-Bacon, 2003).

This fundamentalism of Finn and the right-wing leaders of the United States in the first decade of the twenty-first century, frightens the world in ways that many Americans are only beginning to understand. People around the world are baffled that such scholars seem to believe that there is only one objective history of the world and such a chronicle is constructed from an American point of view. "Do they not understand the arrogance and ethnocentrism of such a perspective?" scholars from Korea, Spain, Germany, Brazil, Turkey, Mexico, and many other countries ask me as I travel around the world. They are profoundly disturbed by where this recovery project may take America and how it may shape U.S. relations with the rest of the world. When Lynne Cheney (2002) argues in her chapter in the Fordham report that in response to 9/11, American teachers need to teach about traditional documents and great speeches of American history—all of which should be in the social studies curriculum, we all agree—she misses some important dimensions of such a pedagogy.

While it is necessary to teach about the historical ideals of the United States it is also important to study the struggles to *enact* such principles in both American domestic and foreign policy. The devil is in the details of these struggles, endeavors marked by profound successes and profound failures. Contrary to the party line of Finn and his compatriots, the study of the failures is not anti-American but a celebration of one of the central ideals of American democracy. As has been argued by many since the emergence of democratic impulses in a variety of cultures around the world, a society is democratic to the degree that it allows for self-criticism. Self-criticism does not seem to occupy a very high rung on the Fordham ladder of democratic values or in the right-wing politics of knowledge in general. It is indoctrination that seems to be at odds with such democratic principles.

Thus, the recovery of positivism, its accompanying ethnocentrism, and its certainty concerning its ability to produce universal truth harbor profound consequences for all inhabitants of the planet. Positivism is one of many ways of seeing the world that achieved and now is re-achieving a hegemonic position. Positivism's reductionism holds no intrinsic, transcultural claim to the truth. Numerous local knowledges help us better understand the world and even ourselves. As we know, however, right-wing education does not sanction a curriculum that allows American students to view their country as others around the world see it. In right-wing schools, students are left unaware that many peoples around the world view American academic knowledges about them as a form of violence.

Over the last century, American scholars have produced information about diverse societies and peoples around the world for the purpose of more effectively exploiting their land, labor, and natural resources—for example, oil. In the Vietnam War, for example, the U.S. military used anthropological studies of particular indigenous peoples in Vietnam for the purpose of winning their trust so they could be manipulated to support American political and military needs. Positivistic knowledge speaks *about* the "culturally different" but not *to* the culturally different. As millions of people around the world have come to understand, the alleged neutrality of such knowledge works for American interests—not the needs of the people under scrutiny (Sponsel, 1992). Thus, positivistic objectivism wrapped in the flag of hard science and intellectual rigor tends to exploit the less powerful, as it separates researchers from the world and its people—especially around issues of ethics, power, and emotion. Thus, rigor in a positivistic context involves being separated from the world. The less we feel the pain of the oppressed, smell death in a war zone, understand what it's like to be disempowered, are to be thought of as inferior and stupid, positivism asserts, the better we become at producing the truth.

In this context, we fall into a logic of fragmentation. Such a logic compels us in the name of objectivity to fragment content and context, information and value questions, and knower and what is to be known (Hayles, 1996). This fragmentation has profound concrete consequences. When content is fragmented from context, researchers assess students' intelligence as if their background has nothing to do with "how smart they are." A student whose parents are both lawyers has a profoundly different experience with, say, language than a student whose parents dropped out of school in the first year of high school. Such a difference—although it has nothing to do with native intelligence—shows up on an IQ test. Without an understanding of context, the student with uneducated parents is deemed "unintelligent" and is advised not to continue academic pursuits. This is a grave injustice.

Values and information are connected at numerous levels. One example of this relationship might involve the knowledge we produce as educational researchers about questions of racial inequality and its relationship to education. If we do not *value* the effort to understand racism and subvert its impact on students of color, then no information is produced about the topic that can be used to help teachers and educational leaders deal with it. The knower and the known are connected as well. If we are examining a study of race and education, for example, produced by researchers who use data produced by white supremacist organizations to prove the intellectual inferiority of Africans and Latin Americans, we can begin to see that the belief structures of the individuals who produce data (the knowers) are intimately connected to the information they produce (the known). If we do not seek out this connection between the knower and the known, we will probably be severely mislead by the knowledge we consume. Indeed, we must be attuned to positivism's logic of fragmentation.

One can quickly discern that people around the world could easily be offended by the fragmented knowledge produced by positivism that often operates to position them in an inferior status to white, European people. In this context, we can see why the reassertion of the superiority of positivist ways of producing knowledge is an important aspect of the recovery movement. Central to positivist laboratory methods is the isolation (fragmentation) of a phenomenon from the environment in which it developed. Many indigenous peoples around the globe take a very different approach to research. Many Native Americans, Australian Aboriginals, African tribespeople, and the like have viewed objects of inquiry in a larger holistic context, attempting to understand their relationship to the environments that shaped them. How can we remove them from their environments, many indigenous peoples ask, when doing so would erase the multifaceted forces that are involved in the complex processes at work? Only with an understanding of context, process, and interrelationships can solutions to problems be formulated that encourage the well-being of both human beings and their environmental contexts.

Positivism overtly rejects these types of connections. The evidence-based positivist science of NCLB scoffs at studies of the lived world of student feelings and emotions—not only their intrinsic importance but also their impact on educational performance. Numerous right-wing educational spokespeople have pointed out that schools do not exist to make students feel good, not realizing that the emotional health of children and young people cannot be separated from the learning process. There is no conflict between concern for student well-being and a rigorous and challenging curriculum. In this domain it is important that Americans be sufficiently humble to learn from diverse peoples around the world, to listen to the profound insights they offer us about epistemological matters. Positivistically enculturated Americans often laugh at many indigenous peoples' descriptions of their talking to the trees, the wind, the rocks, and the animals. Such Americans have not been sufficiently sophisticated to understand the complex metaphorical dimensions of such comments.

To converse with the world in this indigenous sense, an individual has to be well educated in the "languages of nature." Such an education is grounded upon learning to view oneself as inseparable from the physical world, to make sure the world of humans, diverse contexts, processes, and natural phenomena are not fragmented. Positivism is intentionally designed to render us oblivious to such subtle languages. All entities of the earth, many Native Americans maintain, have their own way of speaking. Human beings must learn how to listen to them. In its condescension toward such brilliant ways of understanding the world, ethnocentric positivism is guilty of a form of reductionism that dismisses context and interconnectedness.

Positivistic reductionism fails to discern the holographic nature of reality. This holographic effect is grounded on the notion that all parts contain dimensions of the whole. Many contemporary physicists, psychologists, and sociologists speculate that the universe, the mind, and the interaction between the society and the individual cannot be understood outside of this holographic insight. In positivism, fundamental units of reality (or things-in-themselves) are not deemed to contain data about the larger constructs of which they are parts. Not realizing this dynamic, positivist researchers and educators see everything from atoms, bodily organs, brains, individuals, languages, curricula to television as isolated entities—not as things whose meanings can only be appreciated when viewed as parts of larger wholes and higher orders of reality (Woodhouse, 1996). Indeed, what I am concerned with here is nothing less than the quality of the knowledge we produce about the world and how we confront such information in educational contexts. To counter the irrationality of right-wing knowledge work and pedagogy, we must address both the reductionism of uninformed research methods and the quest for new ways of seeing.

In the intersection of these concerns, we uncover new insights into research and knowledge production, new forms of reason that are directly connected to specific contexts, practical forms of analysis that are informed by social theory and the concreteness of lived situations (Fischer, 1998). Understanding non-Western ways of knowing and the epistemologies of indigenous and other marginalized groups within Western societies, we begin to transcend regressive forms of reductionism. We figure out the epistemological shallowness of reductionistic positivist notions that researchers simply produce facts that correspond to external reality, information that is devoid of specific cultural values. Based on such insights, we begin to realize that the right-wing politics of evidence-based knowledge is a house of cards that collapses as we begin to understand these issues. As we appreciate the historical and cultural dimensions of all knowledge, positivistic proclamations of "how things really are" are exposed as the social constructions they really are. With these understandings as valuable parts of our toolkits, we expand the envelope of research, of what we can understand about the world. We are empowered to produce multiple forms of knowledge that can change the world in democratic and egalitarian ways.

We Have the Truth: The Universality of Positivist Knowledge

All knowledge is local. The Western scientific revolution catalyzed by the work of Rene Descartes, Sir Isaac Newton, and Sir Francis Bacon emerging in the seventeenth and eighteenth centuries was a local phenomenon producing particular forms of local knowledge. Initially, what these great philosophers of science were often attempting to do was to explain why craftworkers in a particular vocation could accomplish amazing tasks. Inexorably, the effort to explain such local phenomena transmutated into a larger attempt to produce translocal explanations. The epistemology that emerged from this grander effort has been referred to as Cartesian reductionism. It is characterized by breaking down a phenomenon into separate pieces, and then studying these fragments in isolation from other parts of the process or even the process as a whole. While this epistemological process has produced many innovations in the past, in the twenty-first century, many scholars from around the world have come to see its flaws.

Such defects impede insight into new domains of reality that are essential for movement to new domains of complexity, new insights in physical and social science, cognition, the ethical domain, and education in particular. Reductionistic approaches become less useful as the complexity of physical and social phenomena increases. Because of the innate complexity of the cognitive, ethical, and pedagogical spheres, the usefulness of reductionistic approaches is immediately undermined. The successful application of positivist reductionism is possible only in those physical and social domains amenable to its methods—for example, questions that lend themselves to issues of frequency or statistical relationship.

There are, of course, an infinite number of profoundly important questions of this type. But questions involving how schools or students are performing are typically more complex that such questions can answer satisfactorily. This is, of course, one of the reasons that contemporary right-wing educational policy fails so severely—it is grounded on a view of knowledge that is ill-equipped to deal with the complex issues of education. The friends of a rigorous, humane, and critical education must address the naïveté embedded in the positivistic quest for universal certainty in knowledge production. The universal knowledge of positivism is the data produced by researchers who use the assumptions of their history and culture—the legitimate history and culture that, too, have been universalized for everyone—to produce "objective" categorizations of all that exists. It is in this context that African peoples can be objectively deemed to have average IQs of 75. In this way universal knowledge becomes colonialist, as it assumes specific ways of seeing (Apffel-Marglin, 1995; Ashcroft et al., 1995, Shankar, 1996).

As we think about this positivist tendency to produce universal knowledges, we realize that our notions of multiple perspectives—multilogicality and epistodiversity—become extremely important in dealing with the colonial power such universalism asserts. The multilogicality and epistodiversity demanded in our attempt to counter the ethnocentrism of positivism must always be critically grounded. Critical grounding in this context means that they are constructed with a power literacy that appreciates the unequal power relations between positivist Western and other ways of producing knowledge about social, political, cultural, philosophical, cognitive, and educational domains. In this context, so-called universal knowledges are produced about education—ways of teaching, the purposes of education, how to run a school, and the proper construction of a curriculum. Teachers in a poor urban school, for example, with high percentages of poor, culturally different, and non-English speaking students, argue that the universal validated knowledges used to shape what they should be doing are ineffective in their unique educational setting.

Even though these teachers and other researchers have produced their own knowledges about these concerns, their knowledges are dismissed as not evidence-based and rigorous. Of course, these teachers know that the universal knowledges forced upon them fail to dignify the special problems and the unique needs of students and educators working in such contexts. Thus, the evidence-based science of the right-wing recovery has little connection to the local needs and complexities of particular educational settings. The asymmetrical power relationship between universal positivist knowledge and other methodologies and cultural ways of seeing is profoundly problematic. The perspectives and actions emerging from universal knowledges too often prove to be harmful for the marginalized and disempowered. Students in the poor and culturally diverse urban schools referenced above are many times positioned as incapable of succeeding in academic work by the universal science of education. Once again the privilege of the privileged is justified and the marginalization of the marginalized is confirmed. And the band played on.

Positivistic universal knowledge is quite remote from the school and student environments about which it makes such grand pronouncements—for example, the lowest quality of student academic achievement is found in this school; that school is one of the "dirty dozen" worst schools in New York City. On numerous occasions I have visited one of New York City's dirty dozen of bad schools. The universal pronouncements of failure consistently fail to take into account the brilliant pedagogical work of particular teachers and the stellar academic performance of specific students. These dynamics are swept away in the power of the epistemology of universal pronouncements. This distance, this spatial/conceptual chasm between universal pronouncements and the social, psychological, and educational activities being researched is irrational—a manifestation of the irrationality of positivist rationalism.

This elevation of the positivist researcher's "truth" over the insights of teachers and other observers is a power play, an insight into the power of positivist science. In this context, one of the most important things that many people do not know about education becomes clear: issues of research methods and their relation to the larger politics of knowledge are central dimensions of the twenty-first century world of politics and education. Critical, democratic educators call for methodological reform in this domain—a new critical complex politics of knowledge. Such methodologies in a power literate politics of knowledge would insist on including the local experience of teachers and students in any research design from the beginning. It would insist on the centrality of normative questions throughout the inquiry process (Hess, 1995; Ross, 1996). For example, how are the needs of students, teachers, and communities being incorporated into the design of educational research? Understanding an educational situation from the perspective of needs of such individuals operationalizes the multilogicality and the epistodiversity discussed above.

What we are calling for here is a reconstruction of human and educational science in our multilogical framework. Such a science challenges and redraws the borders of the positivist educational science now being employed by right-wing reformers for their ideological objectives. Multilogical forms of research drawing upon the bricolage engage diverse forms of knowledge production and combine them in synergistic ways to provide thicker and more ethical forms of insight into social, cognitive, and educational activities and their interaction. Just, for example, including diverse types of people in educational research is a major step toward multilogicality and thicker insight. Few positivist studies listen carefully to and take the insights of students deemed by the school to be failures seriously. Some of the most important ideas I have studied about schooling, its problems and solutions to them have come from such individuals. They often see dimensions of the school or curriculum hidden from those who have succeeded in the institution. Yet, because of their "degraded" status, positivist researchers assume they have nothing important to tell us about education. Ideas obtained from such students often stimulate alternate ways of conceptualizing the processes of teaching and learning and lead educators to new levels of pedagogical cognition.

Students, community members, and even teachers, contrary to positivist assumptions are not passive individuals but active agents capable of amazing accomplishments and insight when given the opportunity to speak (Kloppenburg, 1991). The right-wing politics of knowledge relegates them to a passivity that eventuates in forms of pedagogy that position students as *receivers* of certified information, community members as *consumers* of the product of education, and teachers the *regurgitators* of expert-produced knowledge to be delivered to students. But positivist researchers find it repugnant to see practitioners and clients (students) as knowledge producers. As a form of arrogant knowledge production, positivism prepares its scientists to take their rightful place at the top of the intellectual food chain. How dare such plebeians infringe on the territory of the expert. This power hierarchy works to regulate and discipline those plebs at the bottom of the status ladder. In the positivist matrix, experts tell the functionaries what to do and how to do it. Watch as Dr Slavin and the school supervisors hand Ms Diaz, the third grade teacher, her script for her reading lesson. This script was constructed on the basis of the universal pedagogical knowledge produced by positivist experts—how can she go wrong?

Robert Slavin, a positivist researcher and founder of the highly scripted "Success for All" program used in numerous U.S. school systems, tells us that we now have knowledge about teaching strategies and educational programs that are replicable in all settings—another way of saying, universal knowledge. In this epistemological context, innovative teaching strategies and knowledges of practice developed by brilliant individual teachers are diversions on our path to our larger goals. Such practitioner insights blur our view of the correct knowledge gained from experimental studies that tell us once and for all what really works in education. Debate about these studies is not possible in the positivist universe—the knowledge they present is verified and thus beyond reproach. For those readers who have taught for 30 years and beg to differ with some of these findings, in the words of Bill O'Reilly of Fox News fame—just shut up. The truth is now "out there"; just get over your differences with Dr Slavin and his intrepid band of infallibles.

Thus, the universal truths of the positivists make the world in the image of how Western/American dominant groups "know" it. In the contemporary metahistorical context laid out in these first two chapters, they *recover* a world of hierarchies that were perceived to be in danger of destruction by the liberation movements of the last century. As it seeks to sweep the cultural inscriptions on all knowledge under the epistemological rug, universalist positivism produces educational data that is just as much a reflection of the real world as is a scientific description of a lightning bolt. In the faux-humble expression of this monologicality, positivist researchers tell us that they are just describing "nature"—this is the natural state of what we call education. Of course, what they describe as nature is a construction shaped by unanalyzed cultural assumptions and the tacit conceptual frameworks of the present historical moment.

The Transcultural, Transhistorical Knowledge of Positivism: White Data

As I have written elsewhere, knowledge does not age well. If we want a sense of what such an observation means, all we have to do is to look at any medical, sociological, psychological, or educational knowledge produced a century ago. The assumptions and conceptual frameworks on which such information was based being so different that the contemporary ones we are accustomed to are no longer hidden. The expert researchers who produced such universal, transcultural, and transhistorical data are revealed for the fallible mortals that they were. What was offered as universal seems peculiarly local and parochial. Schools of education—not unlike schools of liberal arts and sciences—have to do a better job of preparing researchers to deal with these epistemological issues. Like researchers in all fields, educational researchers can go through a PhD program and never address the epistemological domain (Hellstrom and Wenneberg, 2002; Fleischman et al., 2003; Street, 2003).

Of course, if the proponents of NCLB get their way such researchers will not have to study these issues. In the United States, and the United Kingdom in particular, one who does not employ universalistic positivist methods is unlikely to obtain funding from government grants. Thus, conversation about the civic dimensions of education—its purpose in a democratic society, questions of justice and mobility, the relationship between curriculum and dominant power, and so on—is further undermined. Such a politics of knowledge renders democratic engagement less and less important in both the educational space and in the corporatized mediascape. The ideology of the free market merges with a bizarre partnership of religious fundamentalism—an interesting phenomenon—and positivism to create an explosive right-wing ideology of knowledge that supports dominant power relationships.

A knowledge politics of unabashed self-interest reshapes the epistemological landscape, as indigenous, subjugated, and transgressive forms of information are positioned as a violation of the dominant culture's right to be free of such disturbing perspectives. For example, many conservative white students in college now insist that they have the right to be protected from diverse points of view in their academic experience, especially around issues of race and ethnicity. Television news programs become more and more a form of infotainment that refuses to question dominant power wielders as it pathologizes dissent. The needs of the new American empire are sacrosanct and beyond interrogation in both infotainment and education. The idea that the world presented on television and in education is a particular perspective out of many worldviews is not a part of the public discourse in the middle of the first decade of the twenty-first century. The notion that such a world is a dominant power-inscribed social construction is taboo in the new knowledge order.

The idea that individuals' location in the web of reality helps shape the knowledges they produce or their view of the world is not a part of the culture of positivism. Such a concept is easier for the dominant culture to digest when we discover it in a context of cultural difference than when we encounter it in our own culture or in ourselves. When many Muslim peoples from around the planet, for example, view the American presence in the Islamic world as a form of occupation, the dominant culture dismisses such a notion as an ethnic point of view. Since dominant culture's typically white ethnicity is erased, American proclamations about the good the United States does in the world is positioned simply as the objective truth by the corporatized purveyors of information. This ethnicity-minus-one orientation—everyone is ethnic except the White Anglo-Saxon Protestants (WASPs)—is a key socioepistemological concept in the right-wing politics of knowledge (Sollors, 1995). When ethnicity is otherness, the African American way of seeing cannot be trusted as much as white people's. Whiteness in this context exists in a state of transethnicity and is thus not limited by ethnicity's blinders.

All epistemologies and all research methods emerge from specific historical and cultural contexts. This is why the pursuit of objectivity is a Sisyphean enterprise—we always read the world through glasses cut by our Zeitgeist, language, and culture. The conservative scholars and educators of dominant WASP culture—obviously, many WASPs understand these epistemological dynamics and work hard to address them in a just and fair way—are the grandchildren, children, brothers and sisters, or parents of the agents of colonialism and neocolonialism. Neither do they nor their families have a set of stories about the ravages and indignities of being colonized. Indeed, many of these scholars and educators are overtly antiracist and would never intentionally discriminate against a person of color in any situation. In the epistemological context described here, however, good intentions are not sufficient to deal with the insidious ideological forces at work in the contemporary context. And in an era of John Silbers, Chester Finns, Lynne Cheneys, Rush Limbaughs, Rick Santorums, G. Gordon Liddys, ad infinitum, we cannot count on good intentions.

Even with good intentions white people must make a concerted, thoughtful effort to produce and evaluate knowledges within the frameworks of other knowledge traditions. Admittedly, this is a dangerous act in any field in twenty-first century America. Such a courageous task demands that any rigorous education examine the omnipresent connection between knowledge and its producer—the connection between knower and knower as we have previously labeled it. The commonsensical positivist notion that truth has nothing to do with who produced it can no longer go unchallenged. A central dynamic of a curriculum that purports to be rigorous always revolves around the complex relationship between the knower and the known (Aronowitz, 1996; Scheurich and Young, 1997; Street, 2003). One can easily discern how much more insightful a curriculum grounded on this concept would be, how much better equipped its graduates would be to deal with the diverse world around them than those "regulated" in the ethnocentric, universalistic curriculum of the right-wing politics of knowledge.

In a nation dominated by such a politics of knowledge, it is difficult for well-intentioned people to know from outside this dominant epistemological context. After more than 30 years of teaching, I can testify that I have never encountered young students more devoid of multilogical perspectives on the world around them. Some of the white students who are oppositional in their identities and seek a form of countercultural countenance embrace an extreme right-wing position. Such students have been drawn to white supremacist orientations that are not uncomfortable with fascism. Living and operating within this specific knowledge culture, many students have no idea that school could be structured differently, that different worldviews exist. I have to remind my brilliant doctoral students who are dedicated to teaching an antiracist, anticlass-biased, antisexist, and power literate form of teacher education that they should not be surprised by the cold reaction of many white students. "Your course," I tell them, "may be their first encounter with alternate knowledge forms. Why would they not be suspicious and uncomfortable?"

New forms of science are needed that address the problems of a universalist positivist form of knowledge production. Science is first and foremost a social construction. Even in the realm of physical science, biological being is also a social process. Human beings are connected to the world around them, the universe itself via our mathematical, physical, chemical, and biological structures and none of these dynamics are disconnected in any way from the social, cultural, and historical dimensions of who we are. A science and a curriculum that refuse to integrate these domains fail their intellectual and ethical obligations to their societies. Yet, the rhetoric of NCLB and evidence-based research is grounded on the denial of these social realities. The story they tell is a simple one. Real science is produced by pure reason outside the boundaries of society and culture. It must be this way, the story goes, because any sociocultural infringement on the scientific process corrupts the validity of the research.

It does not seem to disturb positivism that its methods induce it to focus on those dimensions of a phenomenon that best lend themselves to its form of measurement. Those dimensions that are hard for positivists to measure are summarily excluded. Thus, undue emphasis and importance is placed on particular items not because they are so central to making sense of the object of study, but because they fit the research methods so well. Again, we find irrationality in this feature of positivism—a domain that claims to hold the scepter of reason. In educational research, for example, evidence-based research requires some form of standardized testing to function. We can only tell what works if certain "objective" measuring instruments tell us so.

Yet, many people now understand that there is no such thing as an objective instrument of measurement. While there are many, many dimensions of their subjectivity, one in particular involves what they determine to be important outcomes of a learning exercise. When I study particular tests that are used to measure a student's knowledge of, say, American history, what I typically find is that the instrument focuses on rote memorization of particular historical events. The very qualities that make for sophisticated historical thinking are not a part of the test. For example, positivist research instruments on historical learning rarely measure the ability to

1. evaluate the worthiness of historical sources;
2. deal with conflicting sources;
3. interpret and make meaning of raw historical data;
4. construct a historical narrative;
5. understand the discrepancy between the cultural logics of different historical eras and how this complicates our efforts to make sense of the past;
6. appreciate the purposes of historical study;
7. delineate the complexity of the relationship between the past and the present;
8. understand how social change constantly reshapes the way we view the past.

Of course these are just a few of the types of skills needed by scholars of history. Positivist instruments do not measure these abilities because they are complex and do not lend themselves to simple, quantifiable answers. Thus, in an evidence-based curriculum, these important skills are dropped and replaced by lower level, more easily measured historical skills. Such low-level skills typically have to do with regurgitating unproblematized, subjective historical information that glorifies the status quo, the dominant culture. The positivist science of educational measurement in this context does not engage with what many historians would put forth as important forms of historical thinking. Instead, it perpetuates its tendency for detachment, devising external, reductionistic formulas for determining competence (Hess, 1995; Aronowitz, 1996; Scheurich and Young, 1997). In this way positivism shapes the curriculum—epistemology determines what is important about history. What makes it worse is that it is an unexamined epistemology.

What Can We Do about this Knowledge Climate?

Obviously, any understanding of universalist positivism and its impact on politics and education is complex. Concurrently, any effort to address the right-wing politics of knowledge that shapes education in the twenty-first century demands an understanding of such epistemological dynamics. Contrary to many elitists in the academy, I believe everyday people can negotiate and act on the complicated theoretical dimensions of these issues—indeed, many already have. No truly universal way of producing knowledge exists. In this context, what we can do is develop new and productive ways of using multiple ways of seeing, from research methodologies, academic disciplines, social theories, epistemologies, cultural perspectives, ancient historical perspectives to subjugated and indigenous forms of knowledge. This multilogical form of knowledge seeking understands that because of their unique circumstances some groups of people know things that others do not, scholars of particular disciplines understand phenomena that other scholars do not, ad infinitum.

When I was a child my Uncle Paul was a county agricultural agent in rural Virginia who would sometimes take me with him in his visits to local farmers in the hills of southwestern Virginia. One thing that struck me during these visits to the small farms of these unschooled "redneck" farmers was just how much they knew about their land—its growing and grazing conditions in particular. I listened with fascination as the farmers laughed as they told stories about what the agricultural scientists told them to do and how it did not work on their farms. Their ingenious ways of coping with farming problems that failed to fit the universal scientific models of the experts alerted me to their indigenous genius and, though I could not articulate it at the time, the complexity of scientific knowledge and its local application. These wizened farmers knew things that the agricultural scientists did not know—and no doubt the agricultural scientists knew much that the farmers did not.

The lesson I learned about epistemology in reflecting on this childhood experience in relation to my epistemological understandings is not that one form of knowledge is in some simple way superior to the other. The point was that the scientists and the farmers had much to learn from one another. Multilogicality, as I am employing the word in this context, would involve establishing mutually respectful dialog between the scientists and the farmers. The knowledge produced in such a dialog would not be magically true or universal. Instead, the dialogical process would be mutually beneficial as each group came to understand the frames of reference of the other. Informed by these frames of reference, they would better understand how different individuals came to their conclusions about how best to grow corn. Multilogicality would not provide a perfect synthesis or a choice between one body of knowledge or another. It would construct a dialogical context where respective interpretive frameworks were brought into focus. Such enhanced focus would help farmers become smarter farmers and agricultural scientists become smarter agricultural scientists.

Thus, in an educational context, I am not arguing that either the farmers or the agricultural scientists take over agricultural education. The multilogicality I am promoting allows for a wider debate in U.S. schools about the nature of history, science, literacy, mathematics, and many other disciplines. In arguing for such diversity, I am not retreating from particular commitments to racial, class, and gender justice and the need for a literacy of power to help students better assume their roles as agents of democracy. But as I have maintained in these first two chapters, the call for diverse perspectives of multilogicality constitutes a slap in the face to monological right-wing universalism. Critique and an analysis of diverse perspectives is exactly what they do not want in U.S. schools. Any epistemological position that fears synergistic dialog should be questioned about its commitment to democracy.

Positivist universalism, however, rejects such democratic appeals to dialog simply because it produces the truth. This truth, the positivist story goes, will eventually replace all other claims to truth, putting to rest all these fatuous calls for multilogicality. Such a colonialist view of knowledge is a form of epistemological imperialism that catalyzes the work of military, geopolitical, economic, and educational imperialism. Indeed, it is an epistemology for the empire. Local knowledges produced in the confrontation with difference and the synergistic dialog that emerges are the blood enemies of the right-wing politics of knowledge. The individual interpretations and the communities of knowledge and practice that come out of such knowledge work threaten the information hegemony that now exists. This hegemonic knowledge forms the foundation for the standardized schools of the Bush era. Students learn to be learners who get the "correct" answer to all questions and then take on the "correct" social role in corporatized workplaces (Hartman, 2002). Thankfully, many students will resist such a Stepford education. Such resistance, unfortunately, will be viewed as a manifestation of a lack of proper social adjustment and will be positioned as pathological behavior. Resisting blocs of unaccountable power cannot be tolerated in the *recovered* order.

Thus, contemporary education avoids forms teaching and learning that address the most urgent cultural, economic, political, epistemological, social, and environmental problems that confront us. Such a reductionist education seems uninterested in exploring the frontiers of intelligence, in the process, producing students who possess multiple forms of knowledge and can evaluate the strengths and weaknesses of the information that confronts them. A democratic education that is interested in the frontiers of intelligence values highly educated teachers who can evaluate multiple knowledges and the complex, conflicting requirements of everyday schooling in relation to the needs of their students and the exigencies of the larger society. From such deliberations is great teaching generated—brilliant practice is not generated in the parroting of scripted lessons. In such a rigorous analysis, new forms of consciousness are constructed—higher orders of thinking are not cultivated in test-driven, rote memory pedagogies.

In this critical context, teachers and students learn new ways of engaging subject matter (Novick, 1996; Fleury, 2004). They learn that knowledge is more than simply true or false, as they begin to discern the conditions of its construction. In this context, they are engaging in a form of critical constructivist (Kincheloe, 2005) thinking—a way of seeing that values a rigorous understanding of how knowledge is produced and the role that power plays in such production. Such thinking empowers teachers and students to connect academic knowledges, subjugated knowledges, and a literacy of power to the development of their worldviews, ethical sensibilities, civic activities, and an understanding of their selfhoods. Such profoundly important abilities emerge in their exploration of the origins of the knowledges they encounter. This is the power of such pedagogy—its importance transcends the particular knowledges engaged. It is in the *relationship* between self and a rigorous understanding of knowledge and its production that life-affirming and life-changing education takes shape. What is going on in the minds of learners, in the teacher's consciousness, and in the interaction of the community of learners as this relationship takes shape? Compare this process with the rote memory work of the right-wing curriculum.

The right-wing pedagogy provides students with really no reason outside of a standardized test to learn anything. School in such a context is just a silly hoop through which kids must jump on the road to adulthood. Obviously, I empathize with bored students sitting through drill-and-test type classes. Concurrently, I empathize with teachers who feel pressured to teach in such a scripted and proscribed way. What constitutes education is a decision made in a community of learners, around the types of issues raised here—the quest for compelling worldviews, ethical sensibilities that lead to courageous civic actions, and new insights into who we are, how we became that way, and what we want to become. Thus, in the critical pedagogy delineated here, the job of the teacher is to create conditions that let students become learners and researchers. Sometimes such a task demands a brilliant and inspirational lecture about a topic on which the teacher is well informed. At other times it means keeping one's mouth shut for 45 min—some of us more verbose teacher types find the latter undertaking quite taxing. As John Lennon put it, "whatever gets you through the night"—or in this case, through the class.

The notion of multilogicality developed here involves diverse approaches in many domains. There are many teaching methods that can be used to create the conditions that let students become learners and researchers. The last thing that teachers and students need in these pursuits is more rules and regulations from outside politicos. Scholar, researcher teachers who understand these epistemological, political, and pedagogical dynamics can accomplish great things in classrooms that respect their diagnostic and prescriptive abilities. These teachers can construct curricula that address the needs of their particular students in their specific locations. The effort to make all students march in lockstep to the same reductionistic drummer, chanting monological forms of covertly politicized knowledge is an educational future better designed for an authoritarian state than a democratic one. It rests beneath the dignity of the principles employed in the development of American democracy.

Democracy and democratic education are founded on a spirit of inquiry—not on political conditioning. In the convoluted rhetoric of the recovery movement, however, the progressive call for teachers and students as researchers, an inquiry-grounded education, is represented as a form of political indoctrination. Moreover, the right-wing curriculum's inculcation of support for traditional forms of power inequalities and injustice is represented in this pretzel logic as a form of liberation from the "oppression" of multilogicality. Pray with me now the right-wing prayer for a righteous pedagogy: "Lord, deliver us from overbearing educators who offer diverse points of view and allow us to make up our own mind about social, scientific, political, economic, and ethical matters. Deliver them from this democratic evil and bless us with an education that requires that we learn the final truth once and for all. Enable us to smite all forms of multilogicality and those relativistic evildoers who propose that we learn to explore complex questions for ourselves."

Thus, we "evildoers" in the name of multilogicality assert that all curricular content knowledge be offered in the spirit of inquiry. Students in this context are expected to learn how to teach themselves and self-evaluate their progress. This does not mean that the teacher becomes less involved in such student self-management—in fact, the contrary is usually the case. It merely means that teachers develop a new type of relationship with students; teachers become co-learners, co-researchers with them; to be sure the knowledges produced will not be value-free—although the data produced by some teachers and some students will be grounded on different values than other teachers and students. One of the analytical tasks of teachers and students in such a pedagogical context will involve being capable of identifying the value structures that shape knowledges—their own and that of others. Here students and teachers become savvy knowledge workers who use their understanding of the production of knowledge to obtain more power over their lives, more insight into the construction of their own consciousness, and to engage in civic action to democratize the politics of knowledge.

In the right-wing politics of knowledge, schools exist to distribute knowledge. Should not savvy democratic get a little suspicious of an educational effort to distribute information that studiously avoids the production and ideological dimensions of knowledge? I would think that democratic citizens would have several questions about such an effort. Whose knowledge is it? What are its ideological consequences? Who gets to choose it? One of the right-wing criticisms of progressive pedagogies that emphasize analytical abilities similar to the ones promoted here involves the accusation that such educational approaches never teach "content." Such accusations are bogus in the critical orientation championed here, for they assume a binarism between knowledge and analytical skills. Indeed, in our pedagogy, there is no analysis without subject matter and no subject matter without analysis. As stated earlier in this chapter, all curricular knowledge is approached and/or produced in the spirit of inquiry. There is no a priori assumption of its veracity.

Movin' Out: On to a New World of Teaching and Learning

Thus, we are advocating a form of meta-learning, defined simply as a process of always monitoring the assumptions, hidden rules, and expectations of the learning process. In this form of learning, teachers and students stand back from the process and evaluate the benefits and liabilities of engaging in particular forms of learning. Concurrently, they imagine new and better ways of constructing the learning process. Teachers and students work hard to develop criteria for judging the value and usefulness of their learning. Using such criteria, they explore diverse disciplines of knowledge and, of course, subjugated knowledges that can be used to vivify and enhance certain learning projects on which they are focused. Even in a repressive right-wing curriculum, critical teachers can turn such indoctrination pedagogies on themselves, simply by bringing these questions about knowledge production and meta-learning into the classroom. It is amazing what can happen when a teacher helps students research the official curriculum being forced upon them (Schubert, 1998; Ohanian, 1999; Lester, 2001).

As students become more sophisticated in this meta-learning, they become rigorous students of disciplines. In the right-wing politics of knowledge, such efforts are irrelevant—just give the teachers and students information, then monitor how well they teach it and learn it. Teachers and students become analysts of the discourses of disciplines. A discourse is defined as a constellation of hidden historical rules that govern what can be and cannot be said and who can speak and who must listen. Discursive practices are present in technical processes, institutions, modes of behavior, and, of course, disciplines of knowledge. Discourses shape how we operate in the world as human agents, construct our consciousness, and what we consider true. Teachers and students who study disciplines of knowledge in this meta-analytical context identify the discourses that have shaped the discipline's dominant ways of collecting data, interpreting (both consciously and unconsciously) information, constructing narratives, and evaluating and critiquing scholarship.

An awareness of these dynamics creates a meta-consciousness of the ways unexamined assumptions shape both research in the discipline in general and the validated knowledges that emerge in this process (Madison, 1988; Gee et al., 1996). This is the type of rigorous learning that needs to be taking place in our schools. Teachers and students with this historically informed discursive understanding of a discipline know a field in the context of how it has been used in the world and who used it and for what purposes. In such a context, learners understand diverse aspects of a discipline, in the process coming to understand the cognitive, epistemological, political, and pedagogical limitations of a field of study.

Such learning constitutes a profound act of rigor in the struggle for intellectual development. Indeed, what teachers and students are studying here involves a discipline's rules of construction. Always aware of the complexity permeating knowledge production, our critical pedagogy understands that in order to survive, disciplines had to embrace particular features and structures at specific historical points in their development. Often such dimensions live on in new epochs of disciplinary history, serving no pragmatic purpose other than to fulfill the demands of unconscious tradition. When the teachers' and students' historical discursive study uncovers such anachronistic dynamics, they can be challenged as part of the effort to facilitate more rigorous and pragmatic scholarship, pedagogy, and learning. In this context, critical teachers and students come to understand how certain moral positions, particular modes of public behavior, specific systems of belief, and dominant ideologies were produced by certain disciplinary discourses and knowledge traditions. Such insights are profoundly liberating to teachers and students as they enable such scholars to make better-informed decisions about how they fashion their personal, moral, vocational, and civic lives. So often it is in these ignored discursive practices of disciplines that scholars come to grasp the way power operates to oppress and regulate. Learning to engage in such forms of analysis is a central ingredient of the antidote needed to subvert the right-wing educational agenda.

Such meta-understandings of how knowledge is produced, cultural ways of seeing are constructed, and the status quo is molded are central dimensions of becoming an educated person. Moreover, such insights are keys to our ability to move beyond the oppressive present, to escape the ghosts of history that undermine our efforts for just and humane action. In this educational context, the value of such an education depends on what it empowers us to do. Teachers and students in a critical education create a form of conceptual distance between themselves and their learning that allows them to evaluate how their learning does or does not empower them. Does the learning help them shape worthy personal goals and then facilitate their accomplishment? Does it enable them to work better with other people in collective efforts for personal and social change?

Thus, learning in the critical sense offered here causes us to experience the world in qualitatively different ways. In such an education we discern a change in our relationship to the world—a change that holds profound social, epistemological, political, ethical, aesthetic, and ontological (having to do with our being in the world, who we are) consequences. Thus, critical teachers do not simply deliver knowledge to students but reflect on the types of situations in which students encounter knowledge. If all knowledge is situated—meaning that it emerges in particular contexts and is intimately connected to these contexts—then the context in which teachers and students work with and produce knowledge is an important aspect of the acts of teaching and learning (Hoban and Erickson, 1998).

For example, do we encounter knowledge in a way that is connected primarily to preparation for a standardized test or in an effort to better understand a phenomenon that has a profound impact on our life? In my life as a scholar, my situatedness as a rural Tennesseean from the mountains of southern Appalachia has had a profound impact on my relationship to the knowledges I have encountered. In my choice of studies, I choose to write a history thesis and a doctoral dissertation on fundamentalist Christianity and its social, political, and educational effects. My study and research are intimately related to my identity and the deepest realms of my consciousness given my personal interactions with fundamentalism in my childhood. Such situatedness provided me not only profound motivation to study the topic, but also a conceptual matrix, a mattering map on which to position what I was learning. The critical education promoted here seeks to find such dynamics in the lives of all students in the pedagogical effort to make learning meaningful and a source of passion and commitment. This is the central task of a progressive pedagogy.

Thus, learners are ultimately responsible in a critical education for teaching themselves, interpreting, and producing their own knowledges. Regressive critics of progressive pedagogies have been quick to scoff at this proposition, resorting in the process to a misleading representation of what such self-directed learning actually involves. To set the record straight, such a form of learning takes far more work and a higher level of expertise on the part of the teacher than traditional transmission pedagogies. To engage the student in such a process, adept teachers must not only develop deep understandings of students but must also possess high-level scholarly/research skills and a wide body of diverse knowledges. The teacher must be able to exercise profound pedagogical skills that cultivate the student's disposition to develop such self-directed abilities.

In this complex process, such teachers must not only spend countless hours helping students to connect their personal histories to diverse knowledges but also engage the student in developing conceptual frameworks in which to make sense of such connections (Barr and Tagg, 1995). Such frameworks are central to this learning process, for it is around them that the purpose for learning and, for that matter, living takes shape. Without a sense of purpose, the goals of a critical pedagogy cannot be achieved. Indeed, everything in the critical curriculum leads down the yellow brick road to the Emerald City of purpose. When learners act with the benefits of this ever evolving, elastic sense of purpose, they begin to discern the whole of the forest rather than simply the isolated trees. It is an amazing moment in the lives of teachers and students when the epiphany of purpose grabs a student and shakes her very soul. This is the money shot of a critical pedagogy.

Deprofessionalized teachers and a dumbed down curriculum—to employ the device of understatement—do not contribute to the complex requirements of a successful critical pedagogy. The multiple abilities, the multilogical insights required for such a pedagogy demand a rigorous mode of professional education where the expectations are high and the demands are challenging. Learning in this context is not simply an individual activity that takes place inside the head of an isolated student. Instead, it is a nuanced and situated form of interaction that produces knowledges that change who we are. Learning in this critical situated context cannot be separated from ontology—our being. Our identity always exists in relationship to our learning—knowledge and identity are inseparable. After engaging in certain acts of learning, therefore, I can never be the same. With this in mind we come to appreciate that knowledge is not a thing. It is, instead, a relational process that emerges in the intersection of a wide variety of forces. It is a relationship connecting self, other people, power, and the world with particular conceptual frameworks.

Thus, knowledge is more complex, than we originally thought (Larson, 1995; Fenwick, 2000; Reason and Bradbury, 2000; Thomson, 2001). And because it is so complex, it requires a new mode of pedagogy and new understandings of teaching and learning. Such new understandings help us understand the role of knowledge in our individual lives and in the larger social order. We talk about knowledge, build institutions for transmitting it, but rarely pause to consider the impact it makes in the world. These are the substantive dimensions of education in a democratic society. Once we have asked and attempted to answer them, we simply cannot return to the horses and duckies of the right-wing politics of knowledge and the education it constructs. The universe implicit within these regressive and presently dominant forms of epistemology and pedagogy is a simple-minded place. Such a conception, I believe, insults the complexity and even the sanctity of creation. It certainly degrades the human beings that inhabit it. The effort to get beyond such reductionist, libidinally repressed, parochial, unjust, and anti-intellectual orientations is an objective worth fighting for in this bizarre new century.

References

Apffel-Marglin, F. (1995). Development or Decolonization in the Andes? *Interculture: International Journal of Intercultural and Transdisciplinary Research*, 28, 1, pp. 3–17.
Aronowitz, S. (1996). The Politics of the Science Wars. In A. Ross (ed.), *Science Wars*. Durham, NC: Duke University Press.
Ashcroft, B., G. Griffiths, and H. Tiffin, (eds.) (1995). *The Post-Colonial Studies Reader*. New York: Routledge.
Barr, R. and J. Tagg (1995). From Teaching to Learning: A New Paradigm for Undergraduate Education. *Change*, 27, 6.
Bogle, C. (2003). Cuts in Education Funding Will Improve Academic Performance. Honest. http://www.wsws.org/articles/2003/aug2003/educ-a28.shtml (accessed on April 12, 2004).

Cheney, L. (2002). Protecting Our Precious Liberty. Thomas B. Fordham Foundation. *September 11: What Our Children Need to Know*. http://www.edexcellence.net/sept11/september11.pdf (accessed on March 10, 2003).

Coles, G. (2003). Learning to Read and the "W Principle." *Rethinking Schools*, 17, 4. http://www.rethinkingschools.org/archive/17_04/wpri174.shtml (accessed on April 12, 2004).

Fenwick, T. (2000). Experiential Learning in Adult Education: A Comparative Framework. http://www.ualberta.ca/-tfenwick/ext/aeq.htm (accessed on March 10, 2003).

Finn, C. (2002). Introduction. Thomas B. Fordham Foundation. *September 11: What Our Children Need to Know*. http://www.edexcellence.net/ sept11/september11.pdf (accessed on March 10, 2003).

Fischer, F. (1998). Beyond Empiricism: Policy Inquiry in Postpositivist Perspective. *Policy Studies Journal*, 26, 1, pp. 129–146.

Fleischman, S., J. Kohlmoos, and A. Rotherham (2003). From Research to Practice: Moving Beyond the Buzzwords. http://www.nekia.org/pdf/ed_week_commentary.pdf (accessed on April 2, 2004).

Fleury, S. (2004). Critical Consciousness through a Critical Constructivist Pedagogy. In J. Kincheloe and D. Weil (eds.), *Critical Thinking and Learning: An Encyclopedia for Parents and Teachers*, pp. 185–189. Westport, CT: Greenwood.

Foley, A. and R. Voithofer (2003). Bridging the Gap? Reading the No Child Left Behind Act against Educational Technology Discourses. http://www.coe.ohio-state.edu/rvoithofer/papers/nclb.pdf (accessed on July 14, 2005).

Fox, T. (1999). *Defending Access. A Critique of Standards in Higher Education*. Portsmouth, NH: Boynton.

Gadamer, H. (1975). *Truth and Method*. New York: Seabury Press.

Gee, J., G. Hull, and C. Lankshear (1996). *The New Work Order: Behind the Language of the New Captalism*. Boulder, CO: Westview.

Gresson, A. (1995). *The Recovery of Race in America*. Minneapolis, MN: University of Minnesota Press.

Gresson, A. (2004). *America's Atonement: Racial Pain, Recovery Rhetoric, and the Pedagogy of Healing*. New York: Peter Lang.

Grossberg, L. (1992). *We Gotta Get Out of this Place: Popular Conservatism and Postmodern Culture*. New York: Routledge.

Harding, S. (1996). Science Is "Good to Think with." In A. Ross (ed.), *Science Wars*. Durham, NC: Duke University Press.

Hartman, A. (2002). Envisioning Schools beyond Liberal and Market Ideologies. *Z Magazine*, 15, 7. http://www.zmag.org/amag/articles/julang02hartman.html (accessed on July 14, 2005).

Hayles, N. (1996). Consolidating the Canon. In A. Ross (ed.), *Science Wars*. Durham, NC: Duke University Press.

Hellstrom, T. and S. Wenneberg (2002). The "Discipline" of Post-Academic Science: Reconstructing the Paradigmatic Foundations of a Virtual

Research Institute. http://www.cbs.dk/departments (accessed on July 14, 2005).
Herrnstein, R. and C. Murray (1994). *The Bell Curve: Intelligence and Class Structure in American Life*. New York: Free Press.
Hess, D. (1995). *Science and Technology in a Multicultural World: The Cultural Politics of Facts and Artifacts*. New York: Columbia University Press.
Hoban, G. and G. Erickson (1998). Frameworks for Sustaining Professional Learning. Paper Presented at the Australasian Science Education Research Association, Darwin, Australia.
Horn, R. and J. Kincheloe (eds.) (2001). *American Standards: Quality Education in a Complex World—the Texas Case*. New York: Peter Lang.
Karp, S. (2002). Let Them Eat Tests. *Rethinking Schools*. http://www.rethinkingschools.org/special_reports/bushplan/eat164.shtml (accessed on August 2, 2003).
Keith, N. and N. Keith (1993). Education Development and the Rebuilding of Urban Community. Paper Presented to the Annual Conference of the Association for the Advancement of Research, Policy, and Development in the Third World. Cairo, Egypt.
Kincheloe, J. (2001). Describing the Bricolage: Conceptualizing a New Rigor in Qualitative Research. *Qualitative Inquiry*, 7, 6, pp. 679–692.
Kincheloe, J. (2005). *Critical Constructivism*. New York: Peter Lang.
Kincheloe, J. and K. Berry (2004). *Rigour and Complexity in Educational Research: Conceptualizing the Bricolage*. London: Open University Press.
Kincheloe, J., S. Steinberg, N. Rodriguez, and R. Chennault (1998). *White Reign: Deploying Whiteness in America*. New York: St. Martin's Press.
Kitts, L. (2004). Keep Special Interests out of America's Classrooms. http://www.hoodrivernews.com/lifestyle%20stories/067%20special%20interest%20opinion.htm (accessed on August 2, 2003).
Kloppenburg, J. (1991). Social Theory and the De/reconstruction of Agricultural Science: Local Knowledge for an Alternative Agriculture. *Rural Sociology*, 56, 4, pp. 519–548.
Larson, A. (1995). Technology Education in Teacher Preparation: Perspectives from a Teacher Education Program. Paper Presented at American Educational Studies Association, Cleveland, Ohio.
Lather, P. (2003). This IS Your Father's Paradigm: Government Intrusion and the Case of Qualitative Research in Education. http://www.coe.ohio-state.edu/plather/ (accessed on July 14, 2005).
Lester, S. (2001). Learning for the Twenty-First Century. In J. Kincheloe and D. Weil (eds.), *Standards and Schooling in the U.S.: An Encyclopedia*, 3 vols. Santa Barbara, CA: ABC-Clio.
Madison, G. (1988). *The Hermeneutics of Postmodernity: Figures and Themes*. Bloomington, IN: University of Indiana Press.
McNeil, L. (2000). *Contradictions of School Reform: Educational Costs of Standardized Testing*. New York: Routledge.

The Memory Hole (2002). Department of Education to Delete Years of Research from Its Website. http://www.thememoryhole.org/index.htm (accessed on July 14, 2005).

Novick, R. (1996). Actual Schools, Possible Practices: New Directions in Professional Development. *Education Policy Analysis Archives*, 4, 14.

Ohanian, S. (1999). *One Size Fits Few: The Folly of Educational Standards.* Portsmouth, NH: Heinemann.

OMB Watch (2002). U.S. Department of Education Website Drops Record Number of Electronic Files. http://www.ombwatch.org/article/articleview/1130/1/96?topicid = 1 (accessed on July 14, 2005).

Reason, P. and H. Bradbury (eds.) (2000). Introduction: Inquiry and Participation in Search of a World Worthy of Human Aspiration. *Handbook of Action Research: Participative Inquiry and Practice.* Thousand Oakes, CA: Sage.

Rodriguez, N. and L. Villaverde (eds.) (2000). *Dismantling White Privilege.* New York: Peter Lang.

Ross, A. (1996). Introduction. In A. Ross (ed.), *Science Wars.* Durham, NC: Duke University Press.

Sarder, Z. (1999). *Orientalism.* Philadelphia: Open University Press.

Scheurich, J. and M. Young (1997). Coloring Epistemologies: Are Our Research Epistemologies Racially Biased? *Educational Researcher*, 26, 4, pp. 4–16.

Schubert, W. (1998). Toward Constructivist Teacher Education for Elementary Schools in the Twenty-First Century: A Framework for Decision-Making. my.netian.com/~yhhknue/coned19.htm (accessed on January 7, 1998).

Shankar, D. (1996). The Epistemology of the Indigenous Medical Knowledge Systems of India. *Indigenous Knowledge and Development Monitor*, 4, 3, pp. 13–14.

Sollors, W. (1995). Who Is Ethnic? In B. Ashcroft, G. Griffiths, and H. Tiffin (eds.), *The Post-Colonial Studies Reader.* New York: Routledge.

Sponsel, L. (1992). Information and Asymmetry and the Democratization of Anthropology. *Human Organization*, 51, 3, pp. 299–301.

Street, B. (2003). What's "New" in New Literacy Studies? Critical Approaches to Literacy in Theory and Practice. *Current Issues in Comparative Education*, 5, 2.

Sudetic, C. (2002). The Betrayal of Basra. *Utne Reader*, 110, pp. 45–49.

Thayer-Bacon, B. (2003). *Relational "(E)pistemologies."* New York: Peter Lang.

Thomson, C. (2001). Massification, Distance Learning and Quality in INSET Teacher Education: A Challenging Relationship. http://www.ru.ac.za/academic/adc/papers/Thomson.htm (accessed on January 10, 2001).

Van Manen, M. (1991). *Researching Lived Experience.* Albany, NY: State University Press of New York.

Woodhouse, M. (1996). *Paradigm Wars: Worldviews for a New Age.* Berkeley, CA: Frog.

Chapter 3

What You Don't Know about Standards

Raymond A. Horn, Jr.

Surprisingly, many parents, teachers, and other individuals are not aware of the great impact and deep implications of educational standards, and the evaluation of those standards on individual children and on society in general. This lack of awareness is surprising because of the constant and pervasive evaluation that all members of a school community experience, and because of the short-term and long-term effects of this evaluation on individuals and on the community. Perhaps one reason for this lack of awareness and understanding of the implications of standards and evaluations is the simplistic and superficial presentation of standards and evaluations by those in control of educational policy. In other words, what appears to be a cut and dried aspect of educational policy is actually a very complex and political issue. The political and complex nature of educational standards and evaluations becomes apparent when questions such as the following are asked:

1. What are the different types of standards that can be used to measure student progress?
2. What are the different types of evaluations that can be used to measure the attainment of the standards?
3. What are the purposes to be achieved by the different types of standards and evaluations, and how do these purposes align with the purpose of schooling that we want to foster?
4. What are the consequences of the different types of standards and evaluations for children, parents, community, business and industry, and our country?
5. How should we define student progress in achieving the standards?

6. How much of our children's total learning experience in their school environment should be evaluated against predetermined standards?
7. Who should decide which aspects of our children's total learning experience are more important than others?
8. Who should determine the nature of the standards and evaluations?
9. What is the relationship between the different types of standards and evaluations and their effect on the formation of our children's identity?

Questions such as these require us to go beyond our superficial understanding of standards and evaluations and engage in the deep and hidden consequences of the kind of standards and evaluations that we use and of the specific aspects of our children's school experience that we desire to measure. To explore these deep and hidden consequences, this chapter identifies and discusses the different types of standards and evaluations, and how these types align with different views on the purposes of American education.

Types of Standards

Due to the way standards are presented to the public by the federal and state governments, much of the public believes that there is a political and professional consensus on what should be used as standards for each discipline. In reality, this is not the case. Currently, the standards that are being mandated by many states are technical standards rather than standards of complexity. All standards are simply someone's expectation of what students are to learn. The important point in the recognition that there are different types of standards lies in the phrase "someone's expectations." Expectations of student learning are viewed differently by business people, politicians, educational traditionalists, and educational reconceptualists, as well as by people of different races, cultures, social classes, and places. Each of these different individuals and groups desire to promote their own values and beliefs through the education process. In order to understand the effects of different types of standards, it is necessary to know the values and beliefs of the "someone" who created the standards because those same values and beliefs will be contained within the standards. Knowing these values and beliefs is critical because all types of standards affect the meaning of knowledge, the process of obtaining knowledge, the complexity of knowledge, and the authority of various types of knowledge. Also, they affect teaching as well as learning, and the values, attitudes, and beliefs of the students. Standards and how they are evaluated are, like all things, of a political nature designed to achieve political purposes. There are many terms used to label standards. Content standards refer to what a student should *know* at the end of a course, and are usually focused on the factual information of a discipline. Performance standards identify what a student should be able to *do* by the completion of the course or activity, and usually involves the development of skills. Also, benchmarks are standards that are *samples* of what a student should be able to do at a specific level of development. All of these types of standards can be found within two very different general categories of standards—technical standards and standards of complexity. To what purpose content standards, performance standards, and benchmarks are employed is the distinction between technical standards and standards of complexity.

Technical Standards

Basically, technical standards are reductionist, fragmented, and individualistic expectations of student learning (Kincheloe, 2001a, 2001b). What this means is that the holistic and interconnectedness of knowledge is broken apart into isolated segments or fragments. Breaking apart the whole changes the meaning of the parts and destroys the more complex synergetic understanding of a situation or event. The complexity of meaning that is attained through an analysis of the whole context of a situation or event is lost when the knowledge to be learned or the analytical process used in the learning is restricted to specific information. In technical standards, facts stand alone in meaning or are tightly connected with other facts in order to produce a predetermined meaning that becomes the only correct answer on the test. Other possible meanings or answers that may also be correct in an expanded or different context are positioned as incorrect answers.

One benefit of this control of meaning by reducing knowledge to discrete fragments is the ease in comparing students' abilities in producing the correct answers. The most important attribute of technical standards is the ease in which they can be standardized, objectively tested, converted to statistical data, and used to rank and sort students. An additional benefit of technical standards is that when combined with a standardized test, the full responsibility for learning can be focused on specific individuals such as students, teachers, and administrators. This focus on individual responsibility rather than on shared collective responsibility may provide a smokescreen that hides the more powerful influences on student achievement such as inequitable and inadequate funding, poverty, lack of preschool programs, class size, the scarcity of minority educators, and the political agendas of self-interest groups. Interest groups whose goals are to channel public tax money into parochial schools, private schools, and schools turned over to businesses are aided in their efforts when individual students and schools fail to achieve the test scores mandated by these interests.

In a technical standards environment, students obtain knowledge through transmissional pedagogy, and, within this type of pedagogy or instruction, the skills that students use to acquire and process knowledge are tightly controlled. Transmissional pedagogy is a form of teaching in which the teacher passes or transmits knowledge directly to the student who in turn passively receives the correct and appropriate knowledge. Paulo Freire (1996) called this type of instruction, banking education in this analogy, the teacher has control of the correct, and therefore valuable, knowledge and deposits this knowledge in the minds of the students. The most efficient transmissional instructional strategies include direct instruction techniques such as teacher lecture, reliance on textbook information, worksheets, audio-visual presentations, repetition, and tightly controlled student–student interaction activities. One of the most important roles of the teacher is to make sure that the correct knowledge is not corrupted during the transmission to the student. In technical standard environments in which a standardized test is used to assess student learning, the most efficient instructional technique is to teach to the test.

Also, in relation to the process of students obtaining knowledge, technical standards can utilize all levels of thinking skills, such as recalling, comprehending, applying, analyzing, synthesizing, and evaluating knowledge. However, because the end result of the students' critical thinking must fall within the boundaries of a predetermined correct answer, these thinking skills are designed to be controlled and restricted activities that will result in attainment of the correct answer. Control of these thinking skills is achieved by controlling the conditions of the problem, and by requiring students to follow a specific procedure for the thinking skill that is being utilized.

Concerning the complexity of knowledge, technical standards require the categorization of knowledge into simple binaries of correct/incorrect, right/wrong, appropriate/inappropriate, and authoritative/nonauthoritative classifications. This categorization of knowledge assigns different degrees of value to the knowledge within the categories. One example of this differentiation of value is the identification of appropriate knowledge for students as listed in

E. D. Hirsch's *Cultural Literacy* (1988). Consequently, knowledge that is not listed by Hirsch or that disagrees with Hirsch's interpretation is valued as incorrect, inappropriate, or irrelevant. In technical standards, student discovery and interrogation of the deep and hidden patterns found in a more complex representation of knowledge are de-emphasized in favor of the rote learning of predetermined interpretations constructed by those who control the standards. In fact, student discovery and critical interrogation of the broader contexts in which all knowledge is nested is viewed as a threat to the proposed correctness and authority of the technical standards.

The often-used phrase that knowledge is power not only implies that the attainment and control of knowledge is important but also that some knowledge speaks with more authority than other knowledge. Technical standards are based upon the supposition that knowledge deemed as official or formal carries more weight or authority than unofficial or informal knowledge. The knowledge contained in technical standards is posed as fixed, unchanging, and stable knowledge that exists separately from human interpretation. In other words, a fact is simply a fact and cannot be contested. In this case, knowledge is seen as value-neutral, or as information that is unaffected by human social, economic, cultural, and political conditions and desires. In this case, officially determined knowledge cannot be contested. The correct answer is, indeed, always the correct answer. Often, the information contained in technical standards is posed as scientifically derived knowledge instead of as knowledge that actually represents the mainstream interpretations of the dominant culture. Conversely, knowledge that does not conform to the mainstream interpretation is deemed unscientific and, therefore, lacks the authority that is attached to the mainstream interpretation. Indigenous knowledge, or the knowledge that comes from minority (such as racial, ethnic, or lifestyle) and subordinate cultures (in an educational context such as teachers), is devalued.

As previously mentioned, certain modes of teaching and learning are more appropriate for technical standards. Teacher-centered direct instruction facilitates student learning of the information included within the technical standards. This type of pedagogy directly affects the role of the teacher and the student in the learning process. In a technical standards environment, teaching as a technical and mechanical process is more appropriate than teaching as an art and as a craft. In this context, teachers become deskilled technicians. Deskilling is when the autonomy of the teacher to make decisions is restricted resulting in a loss of general pedagogical and interdisciplinary skills. Instead of being an educational generalist who has a diversity of skills and knowledge, the deskilled teacher is a specialist with few very well-defined and narrow skills and knowledge. Deskilling takes place when teachers are when required to follow prescribed standardized lessons, and when required to use "teacher proof" instructional materials. The deskilling of teachers guarantees that the correct knowledge will be efficiently transmitted to the students. Teacher creativity and responsiveness to individual student needs, abilities, and individual interpretations of knowledge are viewed as uncontrolled variables that may subvert or corrupt the official interpretation and ultimately the correct answer required by the standardized test.

Since all knowledge is connected to values, beliefs, and opinions, all types of standards promote different views about what is to be valued, what is to be believed, and what opinions individuals should hold. Technical standards represent the values, beliefs, and opinions of the individuals or special interests that constructed the standards. Teachers who employ a creative diversity of instructional techniques open the door to student constructions of values, beliefs, and opinions that may not align with the canon, or the official values, beliefs, and opinions that are intended to be reproduced by the required standards. All standards are part of a socialization process that attempts to perpetuate or construct a specific type of citizen and subsequently a specific type of society. Technical standards that tightly control how knowledge is produced are designed to reproduce a society that mirrors the values of those who constructed the technical standards. The current standards and accountability reforms seek to reproduce a society that is characterized by the values and needs of a white, Western European and male, dominated market economy.

Standards of Complexity

Technical standards are in sharp contrast to standards of complexity (Kincheloe, 2001a, 2001b). The only commonalities between the two are that both affect the meaning of knowledge, the process of obtaining knowledge, the complexity of knowledge, and the authority of various types of knowledge. And, they also affect teaching as well as learning, and the values, attitudes, and beliefs of students.

Basically, standards of complexity are holistic, interconnected, and recognize the social, cultural, and historical context of all knowledge, teaching, and human relationships. What this means is that standards of complexity recognize that the individuals within a culture, subgroup, or society construct knowledge socially. In order to understand the broader and deeper meanings of situations, events, and ideas, this recognition then requires the inclusion of the social, economic, political, cultural, and historical influences and interpretations on this knowledge in the learning of the knowledge. Consequently, standards of complexity are not isolated facts but an interconnected body of information whose meaning changes as the sociocultural environment changes. In this case, standards for any discipline represent information that is not reduced to parts but always part of a whole. In contrast to technical standards, meaning is not fixed but is relative to the context in which it currently resides. As context changes, the meaning of the knowledge changes. In highly restricted learning scenarios (such as $2 + 2 = 4$, who was the first president of the United States, and the like), meaning is stable and fixed for all students. However, when the learning scenario becomes part of a real life context, how the fact is used, the purpose of that use, and the sociocultural implications of its use present the possibility of conflicting interpretations and conclusions. In history, one person's freedom to act may well be interpreted by another as an act of oppression. As contextual complexity increases, the ability to insist on one correct meaning becomes arduous. This lack of control over meaning (such as not being able to insist that there is one correct answer) negates the benefit that technical standards enjoy—the ability to compare students based on their ability to arrive at one correct answer. Consequently, standards of complexity do not easily lend themselves to standardized objective tests that can generate statistical data that can be used to rank and sort students.

In a standards-of-complexity learning environment, students obtain knowledge through a transformational pedagogy. In contrast to a transmissional pedagogy, transformational teaching and learning facilitates student construction of knowledge rather than student reception of knowledge constructed by others. Also, since transformational pedagogy recognizes the sociocultural nature of teaching and learning, students learn to work with others to acquire a critical understanding of the effects of knowledge on all individuals. To accomplish transformational learning, efficiency of instruction is redefined because of the diversity of instructional methods that must be employed. Teachers efficiently utilize a plethora of teacher-centered and student-centered techniques so that knowledge can be examined and constructed within its broader sociocultural context. Because of the awareness of the social construction of knowledge and the desire to create transformational learners, the instructional foundation of transformational pedagogy includes cooperative, collaborative, and team problem posing and problem solving. Student initiative and creativity is valued and encouraged.

Like technical standards, standards of complexity utilize all levels of thinking skills, such as recalling, comprehending, applying, analyzing, synthesizing, and evaluating knowledge. However, higher order thinking required by standards of complexity is different in significant ways. How Lauren Resnick (1987) captures the characteristics of higher order thinking resonates with its use in standards of complexity. Resnick proposes that higher order thinking is complex, often results in multiple solutions, involves individual judgment and interpretation, involves uncertainty about the process employed and the conditions of the problem, involves the student to be self-regulated instead of directed by a teacher or a procedure, and involves considerable effort on the part of the student. Standards of complexity require students to use thinking skills within this complex cognitive context.

Also, because of the sociocultural context in which all knowledge is embedded, a critical component must be added to these thinking skills. This additional criticality requires students to include in their thinking about knowledge additional factors, such as race, gender, social class, and any other human contexts that may be involved with the knowledge under investigation. The inclusion of criticality in the teaching of disciplines such as math, science, English, the social sciences, art, music, and others requires students to recall and comprehend basic knowledge. However, when students analyze, synthesize, and evaluate, they are required to go beyond the mere mechanical application of these skills and see more deeply and broadly how situations, events, and ideas affect individuals of different races, gender, and social classes. In addition, student analysis must include recognition of how the historical context of the situation, event, or idea affects the analysis of the situation. The inclusion of criticality in educational standards increases the complexity of student learning. Therefore, when standards of complexity are the basis for curriculum, the teaching of science, math, social studies, language arts, and other disciplines goes beyond the mere presentation and regurgitation of factual knowledge to a critical exploration of this knowledge in relation to other disciplines. In other words, factual knowledge is not separated from its sociocultural contexts. Students not only learn factual information but also at the same time in the same activity learn about the critical implications and consequences of the information within its sociocultural context. One implication of this type of learning is that since all learning is individually and socially constructed, multiple forms of these thinking skills need to be taught to students. The ability to use different analytical techniques allows students to engage the broader context of situations, events, and ideas. Therefore, there is no one correct way to analyze, but many correct ways to analyze the different parts of an idea's context.

Concerning the complexity of knowledge, standards of complexity are different from technical standards in their contextual complexity and in their cognitive complexity. As described, the standards to be learned include not only the factual and foundational information of the discipline under study, but also the additional layers of sociocultural and historical influence that are actually connected to the information in real life. Standards of complexity are cognitively complex because, to learn this information, students must learn and use multiple techniques of each critical thinking skill, and employ their use with an awareness of criticality. Therefore, standards of complexity are complex because their inherent contextual and cognitive complexities require students to go beyond the superficial understanding of information to an understanding of the deep and hidden aspects of information. Another reason why they are called standards of *complexity* is because the meanings that students construct, or the constructed meanings that students critically interrogate, are not simple or cut and dried because of the diverse and multiple interpretations that arise.

This increased complexity problematizes the assignment of authority to knowledge. If multiple contexts create multiple interpretations, can there be one correct answer? Can one body of knowledge be the correct or appropriate one? Can knowledge be proposed as value-neutral? In a standards-of-complexity environment, the answer to all of these questions is no. Therefore, all knowledge bases have a relative or situational ability to speak with authority. For instance, the authority of scientific and nonscientific knowledge bases is related to the sociocultural and historical contexts in which they are embedded. For example, scientifically determined agricultural techniques may be more appropriate for the locale in which they were developed, but indigenous techniques developed in other areas may be more effective in those areas even though they are not considered scientific. Recognition of the relativity of authority requires students to critically interrogate what is posed as official or formal knowledge. In learning standards of complexity, students learn that authority changes as contexts change.

Concerning what constitutes as appropriate modes of teaching and learning, standards of complexity require teachers and students to expand their capacity as teachers and learners. For teachers, expanded capacity refers to the acquisition of a broad range of pedagogical techniques and diverse knowledge. Instead of losing skills through the deskilling process, teachers must continuously expand their instructional *toolbox*. To attain the level of complexity required in standards-of-complexity learning environments, teachers need to know more not less, and to know more diverse teaching strategies rather than be restricted to standardized teaching formulas. Likewise, through their years of schooling, students need to continuously acquire a diversity of learning strategies and skills so that they can graduate with a full *toolbox* that will accommodate the unforeseen learning situations they will encounter in later life. In this respect, both teachers and students are required to be transformational learners who can transform knowledge into useful and critically appropriate applications. Unlike technical standards, teacher creativity and responsiveness to the needs, abilities, and different interpretations of individual students cannot be contained by official interpretations of knowledge or effectively evaluated by standardized tests.

Since all knowledge is connected to values, beliefs, and opinions, all types of standards promote different views about what is to be valued, what is to be believed, and what opinions individuals should hold. Standards of complexity accept the fact that values, beliefs, and opinions will differ among individuals. However, above all, the critical nature of standards of complexity requires individuals to engage in ongoing critical interrogation of their beliefs and those of others. Simply, critical interrogation implies that when individuals construct knowledge, they must fully understand the sociocultural implications of that knowledge. Sociocultural implications include the effect of that knowledge on different human characteristics such as race, gender, social class, and other characteristics that contribute to human diversity. Standards of complexity are part of a socialization process that attempts to develop individuals who recognize and promote social and cultural diversity through responsible and responsive participation in our democracy. Unlike technical standards that adhere to a melting pot theory that strives to reproduce the values of a dominant culture, standards of complexity adhere to the theory that strives to construct a just, equitable, and caring society that values difference and diversity.

Types of Evaluation

There are many ways to evaluate student progress in attaining educational standards, but the educational purpose and philosophy represented by the different types of standards requires evaluation methods that align with their specific purpose and philosophy. Before discussing how different types of standards align with different types of evaluation, there are some general points that can be made regarding all types of evaluation. For the purposes of this discussion, the terms evaluation and assessment will be synonymous.

The first point involves the purpose of evaluation. When educators talk about the purpose of evaluation, two initial purposes are expressed as formative and summative assessment. Formative assessment involves the appraisal of student knowledge or skills prior to instruction. Though formative assessments can be in many different forms (such as pretests, interviews, review of previous work), they are all used by the teacher in planning instruction, and in gaining a better understanding of the student's current level of development or learning. Summative assessments are those that are conducted after instruction and assess student achievement of the standards. Both formative and summative assessments may occur within all types of evaluation plans.

However, historically, educational evaluation has served other purposes than just diagnosis of student progress. Many educators have written about the ranking and sorting function of evaluation. The purpose of using evaluation as a ranking and sorting mechanism is to aid in the decisions about which students can enter certain grade sections, course tracks, elitist courses and programs, and eventually colleges and universities. Of course, the ranking and sorting of students not only affects the students' positionalities within a school but also their eventual position or social class within the larger society. Ranking and sorting evaluation is a holdover from the factory/industrial influence on education that was prominent in the late 1800s and 1900s. The idea was that education should be run like businesses and industrial organizations. Therefore, since businesses needed to rank and sort human resources into management and labor, schools should also sort students into similar social categories. With the tracks determined through educational evaluation, college track students would become professionals, business track students would become business managers and clerics, vocational students would become skilled workers, and the general track would become unskilled or semiskilled labor.

Another purpose of evaluation is to reward and punish students, teachers, and administrators for deficient levels of student achievement. An example of this is the various plans promoted by the federal government and states to reward schools who perform well on standardized tests and to sanction those that do not. This purpose of evaluation is justified by the alleged need for accountability. How do we know if individual educators, students, and schools have or have not been accountable in their teaching and learning? The answer is by their performance on a standardized test.

This purpose of promoting accountability through rewards and punishments is related to another purpose of evaluation—motivating educators, students, and schools to do what they are supposed to be doing. States that led the initial development of standards and accountability models soon found that merely requiring a standardized test did not sufficiently motivate students to take the test seriously. The answer was to add a high stakes component to the tests.

High stakes tests are standardized tests that have a serious consequence for those who fail. The most common consequences for test failure are not allowing students to graduate regardless of other evidence of academic achievement, not allowing students to pass from one grade to another, closing schools, allowing parents to withdraw their children from the school with deficient test scores, disbanding the school boards and contracting the schools to private companies, or disbanding the school boards and placing the schools under the control of a mayor, council, or some other agency. These are indeed high stakes for individual and collective failure on a standardized test.

An additional purpose of standardized tests is to guarantee the reproduction of specific content, skills, values, attitudes, and beliefs in all students. The term reproduction is used because those who make the tests infuse the test material with their own interpretation of what constitutes appropriate knowledge, values, beliefs, attitudes, and the proper use of skills, and want to see their interpretations continued in the next generation. In this case, the purpose of educational evaluation is to firmly entrench one view of knowledge and values in all students.

Another general point about evaluation deals with the psychological aspects of educational evaluation. Types of evaluations and their outcomes affect the identity formation that occurs in all students. Student performance on educational evaluations informs and mediates the development of a student's identity. Success and failure both affect how students see themselves. Student expectations of their future capability for success as an individual and in future professions are greatly shaped by their performance on educational evaluations. Of course, this influence on the formation of student identity is the fundamental purpose of ranking and sorting evaluations. In addition, students learn other things about life from how they are evaluated in their formative years. If educational evaluation is highly competitive, resulting in winners and losers, the probability is great that students in this kind of environment will transfer this competitiveness to other parts of their lives. Also, those who fear and hate their educational experience because of this competitive punish/reward evaluation system will continue to view education and learning with these emotions in later life. Conversely, educational evaluation systems that are positive and facilitate student growth will foster positive feelings toward education and promote the development of students as lifelong learners.

Related to the points concerning competitive and facilitative evaluation systems and to psychological effects of these systems is the issue of educational equity. Some types of evaluation systems, such as standardized testing systems, are characterized by excessive dropout rates among certain demographic groups. In many states, African American, Hispanics, and economically disadvantaged individuals experience very high dropout rates due to language and cultural biases of the tests. In addition, the individuals in these groups who require long-term and intensive remediation in order to pass the tests acquire only basic skills applicable to the service sector of our economy, and miss out on the educational experiences that can enrich their lives and prepare them to compete on an equal basis with others in their future lives. Many scholars have documented how this intensive focus on basic knowledge and skills perpetuates the current social class hierarchy.

A final general point deals with the false perceptions held by the public about educational evaluation. An example is the issue of grade inflation. Grade inflation is a condition in which too many students attain good grades because of a lack of evaluative rigor on the part of the teachers. Grade inflation has become one of the foundational reasons for instituting technical standards and standardized evaluations. While this is a legitimate concern, other types of standards and authentic assessment systems have produced high levels of achievement in most students. For instance, in a mastery learning and continuous progress system, students are required to incrementally master content and skills at a high level of competence before proceeding to the next level of mastery. High levels of competency are set as goals and, therefore, when mastered by students, result in high levels of success by all students. However, this philosophy is problematic to a ranking and sorting purpose, and to a system that promotes competition among students. If all can achieve, then it is difficult to rank and sort them, and instead of competing with others, students become focused on their own development and are more receptive in cooperating with others. In this example, grade inflation becomes a political tool to promote one purpose and philosophy of education over others.

Evaluating Technical Standards

The inherent nature and purpose of technical standards require standardized tests that are essentially objective and invariably involve high stakes. Objective tests consist of multiple choice questions and, in some cases, open response or essay questions. The need to statistically calculate student scores for comparative purposes requires correct answers that cannot allow individual interpretation. Therefore, essay questions are structured so that individuals who correct the test can consistently determine a correct answer. For purposes of comparison, cut scores are arbitrarily set that determine who passes or fails the test. If too many students pass the test, then rather than upgrading the difficulty of the test, cut scores can be raised to ensure a more balanced pass/fail ratio.

Standardized tests are indeed standardized in that they are the same test for all students and are scored the same. However, students are not standardized and represent different developmental levels and cultural backgrounds. If the test represents the language and culture of a dominant middle class, white culture, students from other cultures will struggle with the test questions. This raises the concern about the validity of the tests. Test validity simply means, does the test measure what it is supposed to measure? If the test is to measure historical, scientific, or mathematical knowledge and if the students cannot understand the questions either because of language or because of culture, is the students' failure a result of the their lack of knowledge in that discipline or a result of their cultural difference? If it is because of cultural difference, or not sleeping well the night before, or not having had breakfast, or not being motivated to take the test seriously, then the test fails to measure content or skill knowledge even if it measures other variables.

In addition, the alleged main strength of standardized tests is their objectivity. An objective test is one that is valid, or measures what it is supposed to measure. First, the objectivity of many standardized tests used by states is not substantiated by rigorous statistical analysis. To claim objectivity, each test question needs to be statistically analyzed to determine its validity and reliability (reliability means that the question will consistently result in the same answer by all who answer the question). Also, it is the objectivity of a test that allows for a comparison of test results among students from year to year. Test results from one year to another are used to determine improvement in students over time. This comparison becomes problematic when test content, format, and cut scores are changed from year to year. Also, many scholars have argued that objectivity is an illusion because what is included in the test and how the test results are interpreted are the subjective interpretations of the test makers and test result interpreters. Also, objectivity is compromised when the test makers must decide what mathematical, scientific, or other disciplinary knowledge is included in the test. Since not all disciplinary knowledge can be tested, only selected information is included in the test. Test makers subjectively decide what information is important enough to be tested. Because standardized tests are designed to measure specific facts and skills, other equally or more important aspects of education are often sacrificed. William Ayers (1993) points out that those aspects that are not measured by standardized tests include "initiative, creativity, imagination, conceptual thinking, curiosity, effort, irony, judgment, commitment, nuance, goodwill, ethical reflection" (p. 116). This is a short list that could be greatly expanded, especially to include more complex use of thinking skills and knowledge.

Another important aspect about the evaluation of technical standards is curriculum alignment. This refers to the adage that the written curriculum must be the same as the taught curriculum, which, in turn, must be the tested curriculum. Curriculum alignment also refers to the coordination of what is taught within and between disciplines from one grade level to another. However, curriculum alignment becomes ethically problematic when instruction is focused solely on the test content. This practice is commonly known as "teaching to the test." This practice is the most efficient way to ensure acceptable pass rates on standardized tests. The problem is that teachers can focus all of their classroom assessments on the type of questions that will be on the test. Also, information that is not tested loses value, and when time is an issue, this information is discarded or displaced from the curriculum. This is an ethical problem in that the test becomes the focus of education instead of the needs of the student and the needs of society. The typical response to the pressures of the test are to lessen or eliminate "nonessential" subjects, such as art, music, physical education, health education, and other non-tested subjects. From an ethical perspective, are students best served by not experiencing these other aspects of life? Also, what are the consequences for students whose abilities are strong in these areas?

In conclusion, three basic questions come to mind when evaluating the effectiveness of the standardized evaluation of technical standards. What have the students learned? Will what they have learned benefit them in their future lives? What purpose and philosophy does this type of evaluation promote?

Evaluating Standards of Complexity

The inherent nature and purpose of standards of complexity requires authentic and multiple assessments within a mastery learning and continuous progress framework. Authentic assessment is evaluation of a student's knowledge and skills within a real-world setting. In contrast, by reducing knowledge and skills to multiple choice and contextually restrictive open response questions, objective testing creates an inauthentic and artificial testing situation that requires the student to merely recall information or perform a skill that is detached from the student's current or future real-world context. Because of the close connection between authentic assessment and the real world, the potential for the student's transfer and use of the knowledge and skill to nonacademic real-world situations is greatly increased. In addition, authentic assessment creates the potential to positively motivate students to learn—resulting in lifelong learners.

Authentic assessments are also called performance tests because students must correctly perform the skill or demonstrate knowledge attainment before moving on to other standards. This sounds similar to students performing on an objective test by providing the correct answer. However, besides the requirement of situating the performance within a real-world context, there are other significant differences. In the real world, demonstrations of knowledge and skill involve the multiple use of skills, the use of different types of knowledge at the same time, problem posing, problem solving, and the involvement of other individuals in the collective solution of the problem. Seldom is an authentic problem resolved by the use of information and skills from one discipline such as mathematics. More often, knowledge and skills from math, science, social studies, language arts, and other disciplines are required to solve the problem because aspects of all of these disciplines are part of the expanded context of the real world problem. Even though performance tests restricted to specific disciplines are appropriate within the learning process, standards of complexity, which reflect the complexity of real-world problems, require equally complex performance tests. In a standards-of-complexity learning environment, more disciplinary restrictive performance tests can be used to shape student learning. However, the interdisciplinary nature and complexity of the evaluations must increase as student capabilities increase and move to a final performance.

Because involvement with others in real-world problem solving invariably occurs, authentic assessments require students to work with others in different capacities in problem solving and in evaluation situations. Authentic assessment involves multiple assessments not only to capture this interpersonal complexity but also to capture a holistic view of student achievement of the standards. The use of individual and group written and oral evaluations are important characteristics of authentic assessment. Grading rubrics are used to better define standards and to more accurately measure student performance. To ensure a detailed and accurate determination of student progress over time, various kinds of portfolios are used. Valid and reliable standardized tests are also appropriate for periodic diagnosis and comparison of students with other students nationwide. And, an attitude is required in which all types of evaluations are viewed as positive feedback in the promotion of student learning. Authentic assessment requires various degrees of a continuous progress view of student evaluation. Simply, continuous progress implies that within a period of time, students progress at their own rate of learning in the mastery of all standards.

Continuous progress is a problematic strategy in schools that emulate a factory system type of education because this strategy requires a flexible school schedule that allows students to progress at their own pace and to receive the appropriately complex remediation that is necessary to achieve at a mastery level. Another problem is that teachers need the knowledge and time flexibility to create additional resources and assessments for each level of performance. Current definitions of the role of the teacher (such as requiring teachers to be technicians instead of high-level professionals), teacher–student ratios, current demands to rank and sort students, and additional educational funding are all impediments to the use of authentic assessments.

Lev Vygotsky proposed the idea that educational evaluation must be designed to understand two aspects of student learning—whether the student learned the knowledge, and the student's *potential for further learning*. Authentic assessment accomplishes both of these goals because, when done correctly, a wealth of evidence is acquired concerning what the student has learned, the nature of the student's potential for further learning, and precisely, what the student needs to learn next. Also, the authentic and multiple nature of this kind of assessment facilitates an understanding of each student's learning regardless of developmental levels and cultural backgrounds. What students from diverse backgrounds learn can be accepted as valid and objective results because of the precise definitions of the knowledge and skills to be learned, and also because of the diversity and quantity of evaluative evidence that is collected over time. Like objective testing, authentic assessment can provide snapshots of student progress and can be used to compare students, but unlike objective testing, it provides an ongoing informed view of the student's continuing progress in mastering the standards of complexity. In an ethical context, student needs are being met, student evaluation is more accurate and therefore fairer, and the needs of a democratic society are being met by the development of an epistemologically and socially competent and critically thinking individual.

Conclusion

Historically, standards and evaluations have functioned as controlling mechanisms through which certain knowledge has been promoted as appropriate and through which students have been ranked, sorted, and moved into different categories, jobs, and subsequently social classes. Many individuals have argued that those who wish to reproduce their own way of life, values, and beliefs have utilized standards and evaluations as a social control mechanism to achieve this reproduction. However, in this new millennium of changing demographics and more complex social, economic, cultural, and political problems, will this instrumental use of education promote or diminish American democracy? The need for standards and evaluations of student learning is a given. However, what we need to address are these questions. When our children graduate, what knowledge, skills, and capacity do we want them to have to engage in future learning? What kind of society will our children construct because of their knowledge, skills, and capacity for future learning? How will we define American democracy? The answers to these questions relate directly to the type of standards and evaluations that are used in American education. Does the education of children for their participation in a complex society require an education guided by standards of complexity and equally complex evaluations of student achievement?

References

Ayers, W. (1993). *To Teach: The Journey of a Teacher*. New York: Teachers College Press.
Freire, P. (1996). *Pedagogy of the Oppressed*. New York: Continuum.
Hirsch Jr., E. D. (1988). *Cultural Literacy: What Every American Needs to Know*. New York: Vintage Books.
Kincheloe, J. L. (2001a). "Goin' home to the Armadillo"; Making Sense of Texas Educational Standards. In R. A. Horn and J. L. Kincheloe (eds.), *American Standards: Quality Education in a Complex World—The Texas Case*, pp. 3–44. New York: Peter Lang.
Kincheloe, J. L. (2001b). See Your Standards and Raise You: Standards of Complexity and the New Rigor in Education. In R. A. Horn and J. L. Kincheloe (eds.), *American Standards: Quality Education in a Complex World—The Texas Case*, pp. 347–368. New York: Peter Lang.
Resnick, L. B. (1987). *Education and Learning to Think*. Washington DC: National Academy Press.

Additional Sources on Standards and Evaluations

Bracey, G. W. (1998). *Put to the Test: An Educator's and Consumer's Guide to Standardized Testing*. Bloomington, IN: Phi Delta Kappa International.
Horn, R. A. (2002). *Understanding Educational Reform: A Reference Handbook*. Denver, CO: ABC-CLIO.
Kincheloe, J. and D. Weil, (eds.) (2001). *Standards and Schooling in the United States: An Encyclopedia Vols. 1, 2, 3*. Denver, CO: ABC-CLIO.

Kohn, A. (2000). *The Case against Standardized Testing: Raising the Scores, ruining the Schools.* Portsmouth, NH: Heinemann.

McNeil, Linda M. (2000). *Contradictions of School Reform: Educational Costs of Standardized Testing.* New York: Routledge.

Meier, D. (2000). *Will Standards Save Public Education?* Boston: Beacon Press.

Wilde, S. (2002). *Testing and Standards: A Brief Encyclopedia.* Portsmouth, NH: Heinemann.

Chapter 4

What You Don't Know about Evaluation

Philip M. Anderson and Judith P. Summerfield

What most of us do not know about evaluation could take a book, a series, or an entire library to remedy. Even most evaluation experts, and we ourselves are not experts in evaluation, will tell you that they do not know everything there is to know about evaluation. But, even the few experts who claim to know everything about evaluation are either fooling themselves or limiting evaluation to a limited set of definitions or formulae. In any case, evaluation is a topic where ignorance and foolishness rule the court. Those of us who teach feel increasingly frustrated and, as a result, end up ceding large areas of assessment to school officials, testing "experts" and the companies who make fortunes out of our students' failures and achievements.

The central problem with knowledge about evaluation is the assumption that statistical and methodological knowledge, that is, technical knowledge, is what we need to be better teachers. Instead, we would argue that fundamental misunderstandings about evaluation lie in social and cultural assumptions, and that many of our everyday classroom assumptions (i.e., our tacit knowledge) about evaluation are imbedded in the history of large-scale standardized testing. The discussion we intend in this chapter takes us to the larger historical and sociocultural picture required for reading or interpreting evaluation as it is currently pursued. It is the human and social foundations of evaluation, not the mathematical and methodological, that is central to employing student-centered evaluation in our classrooms.

We begin with the current trend toward limited, content-based testing, the so-called content standards movement, with its focus on curriculum defined as "essential" knowledge. The emphasis on student achievement through these tests, coupled with the threats of No Child Left Behind to close our schools and "re-assign" the teachers, has gotten our attention. Large-scale content standards testing has moved teaching and learning away from teachers teaching and children learning to teachers being held accountable and the children meeting "rigorous" standards. You can see the problem here. You can observe the problem in classrooms where the evaluation is driven by the test of essential knowledge. Teachers teach to the test—students "learn" information—tests determine whether the students have "learned" anything. Here the harm is the greatest because testing shapes evaluation and evaluation shapes and even determines the curriculum and pedagogy of a classroom.

The current scene is complex, confusing, and troubling: there is an enormous gap between the standardization of curriculum and evaluation and the everyday (every minute?) evaluation that we need to accomplish in our classrooms. Classroom evaluation may appear to be different from standardized tests in important ways, but there must be a relationship between classroom evaluation and the evaluation done through objective testing and large-scale testing ("objective" in this case does not just mean reducing subjective factors to a minimum, but limited to choices of fixed alternatives, e.g., multiple choice questions). Large-scale testing provides a context (or in current corporate education-speak, "benchmark") for our individual work of evaluating students. But clearly, the relationship between the two is not simple or, in some cases, even productive.

The current situation takes us to the great American school debates, the persistent battle over the curriculum—what is to be taught. The current imperative is for the curriculum to represent "essential" or "core" knowledge. The notion that tests must test all students on similar, if not the same, knowledge is central to the content standards testing methodology. This search for essential knowledge (or core curriculum) relies on answering the "what knowledge is of the most worth?" question. The question deals with *values*, that is, a belief that certain knowledge is more valuable than others. In a pluralistic society, this means that some groups' knowledge becomes privileged, that is, gains more official status. The indigenous knowledge of this group, the group members' tacit knowledge (intuitive knowledge), then becomes the most important knowledge. The assumptions of this specific knowledge become the "norm" in the culture. In colonial settings, the knowledge of the less privileged groups is either destroyed (e.g., the Spanish Jesuits burning the Mayan codices) or neglected into marginalized status (see Apple, 1991).

The "what knowledge is of the most worth" question is a characteristic of conservative reform (actually, reactionary) movements. This question is usually asked as a means to reign in change (or progress) within curriculum structures and content. This current essentialist (one culture) movement can be seen as a reaction against multi (many) cultural curriculum—and against a progressive or liberal ideology. The current essential knowledge curriculum appears to be a power play by conservative forces in the society who believe that schools have "strayed" from their purposes of teaching the "basics" and good citizenship. The implications are enormous, since evaluation in this model is intended to drive curriculum and pedagogy.

Not surprisingly, the proponents of the essentialist movement tend to be the same as those who believe that the Supreme Court has strayed from the intention of the original framers of the Constitution in its various court decisions and constitutional amendments over the years. For us to acquiesce to the essentialist curriculum, to adopt the idea that "basics" are the key to knowledge, is to relinquish our jobs as *teachers*. Which knowledge is of the most worth cannot be taken for granted—it requires a prior question about what beliefs we hold not only about curriculum and assessment but also about how we value students and teachers.

What is interesting about most of the current rhetoric is that it is part of a much larger cultural debate that has been raging for over a hundred years. There are several dimensions to the discussion, some focused on political and cultural issues, while others filtered through technical and methodological matters. And, like all the issues surrounding evaluation, there is a set of questions related to the classroom and a set of questions related to the society. As we said before, the larger social questions have a direct impact on classroom curriculum and evaluation.

The societal questions, as they relate to schools, can be equated functionally with college entrance requirements. College entrance examinations in the nineteenth century were based on a *restricted reading list* (core knowledge) developed by individual colleges located primarily in the east. As schools cropped up in the new states and territories of the west, they complained that the schools in the regions where the elite eastern college resided enjoyed an unfair advantage over the schools from outside these communities. The schools asked the colleges to provide a list of expectations for the students to meet that were *uniform*, in other words, a list that was open and public, and that gave their students a fair chance to compete. By the last quarter of the nineteenth century, students expecting to go to college took an examination on the readings from the Uniform Lists (Applebee, 1974).

Over the years, the Uniform Lists provided a common cultural content for school knowledge. Students in Michigan could be expected to read the same books as students in Massachusetts, provided each had the goal of taking the admissions test for Harvard. But, as the country grew and expanded, some argued that the cultural expectations of the eastern colleges should not define the school curriculum. One of the ironies here is that the colleges apparently never intended that the list define core knowledge, but merely be representative in its suggestions for book titles. This is one of the unhappy secrets of recommended lists—they almost always become the required list.

The first decades of the twentieth century were a time of social and cultural transformation—the implicit assumptions of the specified cultural tests as a means to define preparation for college started to seem a bit quaint. Tycoons had established great colleges based on the perceived demands of the new machine age and dedicated to the new industrial world (Carnegie, Stanford, et al.). This was a new world dictated not by the gentlemen's classical education of the past but by the scientific and technical education for the future. The assumptions about what knowledge is of the most worth changed in fundamental ways. The added value of scientific and mathematic knowledge was only part of the change; the new states even questioned the common cultural content of a curriculum dominated by New England authors (Kliebard, 1995).

Several important changes took place in educational thinking during the early twentieth century along with these various cultural changes: William James and various *American* pragmatist philosophers asserted the primacy of experience in learning and knowing; American psychologists oversaw the development of scientific testing, first used for sorting recruits in World War II; and the U.S. Congress legislated, after pressure from unions and social reformers, the requirement of compulsory schooling, with an emphasis on vocational and technical education for the working classes. The focus on *experience* shifted the emphasis on students as passive recipients of knowledge, which was meant to provide and arrange the "furniture of their minds," to students who actively learned by doing. The challenge was to engage the whole child in the process of learning.

Scientific testing was essential to scientific management, the grand new way of managing society after the fall of the old monarchies. Scientific testing assumed certain biological or genetic traits that could be measured at an early age to discover the capable future leaders among the population. One finds various critiques of this approach in celebrated novels such as *Brave New World* by Aldous Huxley and disastrous political experiments during the twentieth century. More recently, the genetic arguments have been seriously reasserted by

Herrnstein and Murray (1994), and forcefully resisted (see Kincheloe et al., 1995). Scientific testing, though, is not the problem—the misuse of scientific testing to oppress difference and to legitimate an unfair system is the problem. IQ became part of our vocabulary, and most of our tests for the meritocracy grow out of the same assumptions about testing that underlie the IQ (a alleged relationship between "mental age" and chronological age). It is part of our uncritical discourse on schooling to accept such constructs as above or below "grade level" based on results garnered from standardized tests that grew out of intelligence and aptitude testing.

Scientific testing also solved the common cultural content problem inherent in the restricted reading list. Scientific testing argued that real potential was not in the content of one's current knowledge, but in one's mental abilities. Mental abilities in these testing situations were generally defined as problem solving, a key skill in a pragmatic society. The tests purportedly tested your skills, rather than your cultural knowledge. Out of this trend—and several experiments showing that the new Scholastic Aptitude Test (SAT) was a better predictor of college success—came the dominance of the SAT (and its Western sibling the ACT) for college admission. Education would not be about possessing specific knowledge, but rather about possessing certain mental skills. Since we were living in a brave new world, a world of discovery and new knowledge, why would a cultural knowledge of the past serve as the basis for schooling?

Compulsory schooling solidified the changes. Suddenly, you could not opt out of the official curriculum any more (unless you could afford a private education). Apprenticeships and other ways into the employment sector were increasingly being regulated, and regulations not only meant tests but also academic credentials. One had to pass tests to get a high school diploma, to be promoted, even to get to graduation. If one hoped to go to college, one needed to pass the college examinations. The scientifically managed society needed a mechanism by which to identify the successful and simultaneously to reward them with additional opportunity.

As we indicated above, the history of schooling in the twentieth century is a continuing debate between the content standards supporters, those who see knowledge as a product to be acquired, and the performance standards supporters, those who see knowledge as a set of skills and demonstrated competencies. The former tend to look at school as separate from the surrounding culture and school knowledge as something to be gotten only in school. The latter see learning and knowledge connected to the lived experience of the students and gained and demonstrated through activity. The former tend to support memorization or "recall" as a measure of understanding where as the latter tend to value strategies and skills as indicators of competence.

All of the crisis points in U.S. education can be played out as a tug of war between these two positions. The early educational psychologists and progressives (performance advocates) were rebelling against the mental disciplinarians (content advocates); the post-Sputnik battle against the progressives resulted in the academic resurgence in the late 1950s and early 1960s; the humanistic revolution of the 1960s and 1970s was a reassertion of humanistic values, and a short attempt at returning to the basics (skills performance) in the 1970s. The current trend in evaluation, active for 20 years, is a result of a reassertion of the content focus.

The most obvious evidence of this trend is in the reports and books of the mid-1980s that announced the principal neoconservative agenda. The *A Nation at Risk* report railed against the "smorgasbord" curriculum, that is, the elective curriculum that had originated in the "liberal" reforms of the late 1960s, and favored a core curriculum of reduced choices and breadth. E. D. Hirsch's *Cultural Literacy* (1986), a highly influential book, was subtitled *What Every American Needs to Know*, and argued strenuously for a defined, and enforced, "common" knowledge. Hirsch attacked the progressive ideal of performance curriculum and testing by arguing that specific cultural content was necessary for educational attainment—he was simultaneously asserting what knowledge was of the most worth. Following shortly thereafter, while both were officials in the U.S. Department of Education, Diane Ravitch and Chester Finn (1988) published the results of their *first* "national test of humanities knowledge" in *What Do Our 17-Year-Olds Know?* One can see the clear content standards focus even in the title.

The reason that there has been no *second* national test of students' humanities knowledge was that the other ideology of the neoconservative movement, the reduction of the federal role in state matters (such as education), took precedence over a national curriculum and a national test. (We should note here that all the countries of Western Europe have national curriculums and national testing systems that sort the students as early as age 11.) Instead, the U.S. Department of Education offered major funded grants to develop "state frameworks" that would serve as templates for state officials to develop curricula appropriate to their states and state-level testing systems. As all teachers and parents know, this work has been accomplished with a vengeance; all U.S. states now have new content standards-based curricula and state-testing apparatuses in place. NCLB has used these testing mechanisms, along with federal benchmarks, to determine the success of schools, and their futures, in those states. These state tests represent the victory of the content model over the performance model. Even during the development of state frameworks, there was considerable discussion of content standards versus performance standards, with Lauren Resnick's performance-based models at the University of Pittsburgh having considerable influence. At one point, the New York City public school system was using Resnick's *performance* standards to develop a local curriculum out of the New York State Department of Education *content* standards frameworks. Needless to say, the effort was conflicted in its purposes and aims, though, to be fair, the New York State curriculum combined content and performance standards in a complex manner.

Probably more troubling, in our current situation, is the use of tests meant to evaluate student learning (learning outcomes) to evaluate teachers and schools. If one reduces all the functions of schooling to student test scores, this makes sense. After all, goes this logic, if teachers and schools are designed to teach students, then the student scores are a measure of the success of the teachers and the school. Then again, if one sees schools and schooling as providing something more than a restricted academic education measured as recall, then the logic is not so obvious. And, as we all know, factors affecting performance on "objective" tests include matters the schools cannot control.

We raise this problem because it opens up the question that is most important here: Testing for what? There are other tests than those represented by the content standards of the schools, each serving a different purpose. There are tests that are meant to make statements about the nation at large, like the National Assessment of Education Progress. There are also tests that carry on the great tradition of psychological testing, the best known being the SAT and ACT test for college admission.

There are various distinctions in evaluation discussions that are quite important, one being the distinction between admissions examinations and exit examinations. The SAT and ACT are admissions tests; they claim to predict success in the first year of college (and nothing else). The SAT leans more toward performance-based testing while the ACT is more content-based, though both basically serve similar purposes based on standardized measures. The tests are also norm-referenced, that is, test-takers are measured against one another and recent populations who have taken the examination.

If you take the SAT and do not perform well, you can still graduate from school—only your choices for college are limited. The state tests are school exit examinations—if you do not pass, you do not receive a high school diploma. State tests are criterion-referenced, that is, you are measured against a predetermined passing score. The state tests are "high stakes" tests for the entire student population; the SAT/ACT scores are only "high stakes" for those hoping to get into a selective college. NCLB has set standards for each specific category of student (race, class, gender, and other categories) and schools must meet the standards. If one layers the NCLB standards onto the state scores, then the scores are high stakes for teachers and schools as well.

Until recently, some states had different examinations for the college-bound, and another for the noncollege-bound. In New York, for example, the Regents Examination reflects a long history of college admissions testing, but there was also a failed experiment in the 1980s and 1990s with a Regents Competency Test for high school exit testing. Part of the new content standards movement is the assertion of a single academic curriculum for all students, with a single measure of achievement to go with the single curriculum. In other words, there is no longer a college track and a noncollege track; ergo, all students are college-bound. New York has kept a modified version of the Regents Examination for its state examination, and all students are being taught the Regents curriculum. (We can report hearing that some of the students who had traditionally been in the non-Regents track in New York High Schools are now tracked into the non-Regents Regents track, but that may only be a rumor.)

One needs to remember that all achievement tests are designed to measure the individual achievement of students. Each student is to be measured against the criterion for knowledge established by the state-testing apparatus. These are achievement tests that measure whether one has successfully navigated the knowledge (and skills and attitudes) dictated by the state in which one attends public school. The norm is not a national norm; one is not measured relative to others' performance, and passing the test is necessary to be credentialed as a high school graduate.

This model has been moved down into the elementary and middle grades as well. Formerly, elementary testing tended to be of the performance type, measured against national norms. Tests such as the Iowa Test of Basic Skills or the California Achievement Tests were the norm. Until recently, no states had developed content-based tests to measure individual elementary student achievement in high stakes fashion. Now, the chancellor of New York City Public Schools (a lawyer by actual training and experience) holds third graders back if they do not pass the third grade test, and is planning to do the same to fifth graders. This is the plan in many states at the fourth and eighth grade levels as well. In every instance, the tests are criterion-referenced, content-based tests. The state tests are based on teaching specific content, not on various skills like critical thinking. The tests test the specified curriculum; in every case the test defines the curriculum.

Another type of test, the National Assessment of Educational Progress (NAEP), situated in the grand tradition of mass, norm-referenced testing, is a *descriptive* test. The NAEP was developed when President Nixon decided he wanted an assessment system in place before he would sign the renewal of the Elementary and Secondary Education Act. The NAEP attempts to measure how U.S. children and adolescents are performing *as a group*, relative to previous groups of students of the same age. Every several years a representative group of students (several thousand out of a potential several million), chosen on all the factors that distinguish students by region and characteristic, take this test of skills in a range of intellectual domains. When one hears that U.S. eighth graders are doing better, or about the same as past counterparts, this is usually reporting from the latest NAEP results. This type of test, a large-scale sample assessment, was one of the approaches considered by the National Educational Goals Panel, created by the National Governor's Association and the National Council on Educational Standards and Testing (charged by Congress) in July 1990, for national standards. Headed by former Colorado governor and recent Los Angeles superintendent Roy Romer, the council recommended "two components, individual assessment and large-scale sample assessment, such as the National Assessment of Educational Progress" (Romer and Fitzgerald, 1996, p. 237). The NAEP is still used as a de facto national test of educational achievement, though it does not appear to carry the weight of the state tests.

The whole business of reporting the results of tests is probably the biggest confusion within the recent fights over testing and assessment. When Ravitch and Finn reported the results of the test that became *What Do Our 17-Year-Olds Know*, they got the most media attention from claiming that a majority of 17 year olds (in 1986) could not name the dates of the American Civil War within 50 years. On the surface, this finding, especially the spread of 50 years, appears shocking. Unreported in the news release, but clear from the published report, was that the question, asking what years the U.S. Civil War was fought, was a multiple choice question with the following choices: (1) 1750–1800, (2) 1800–1850, (3) 1850–1900, (4) 1900–1950. To claim that any one who chose (2) did not know the dates within 50 years strains credulity.

The form of the test is central to what we can conclude from it. If a test is all multiple choice, how can we say anything about the student's writing ability? Similarly, if the test only tests memory and facts, what can we conclude about a student's ability to generate creative thought? If the test is of an individual, how can we conclude anything about that individual's ability to work in the world with others? The battle over curriculum is played out in these areas as well, with the content standards people arguing that content knowledge is a prior condition (e.g., Hirsch's (1986) cultural literacy argument) to other forms of understanding and learning. But learning is much more complex than a linear, logical model of step-by-step acquisition of information.

Given new attempts to define the complexity of learning, there have been several well-documented attempts to both redefine what tests evaluate and introduce changes in the form of tests to reflect different evaluation priorities. Howard Gardner's work (1993) on multiple intelligences questions traditional intelligence (and achievement) tests as merely testing verbal ability, and insists that there are other measures of intelligence that should be part of the cultural imperatives of school testing. Another recent trend, the use of portfolio assessment, in which student-produced and chosen work is evaluated as a whole, has been used for statewide assessment in Vermont. Portfolio assessment provides a way to assess students individually, but with the claim that the assessment is more authentic than standardized testing. Portfolio assessment is, on the surface, more *valid* than multiple choice objective tests, but probably not as *reliable*.

"Validity" and "reliability" are important concepts, as well as statistical constructs, that every teacher should be able to manage. *Reliability* refers to how accurately an instrument measures whatever it is supposed to measure; *validity* refers to how well an instrument measures whatever it measures. There are some excellent books for teachers on testing and assessment; Michael Lorber's *Objectives, Methods, and Evaluation for Secondary Teaching* (1996) is one of the best-written and most sympathetic to teachers' needs (the extensive evaluation discussion is useful for elementary and college teachers as well). In the case of the portfolio, one can see that the validity of the portfolio is its attraction: one is measuring actual student work, and, in most cases, work that the student has chosen to represent his or her best work. The reliability of such a measure is tricky because one is supposed to evaluate each individual on his "authentic" production of work. A standardized, norm-referenced objective test, on the other hand, measures students all performing the same task or knowing the same information. Reliability is enhanced in this setting, but the validity becomes problematic.

Another classic example of the validity/reliability conundrum is the case of the SAT test of grammar and usage being used as a test of writing ability—the test may be quite reliable in testing grammar knowledge, but it is barely a measure of the ability to write (it does not even ask the student to write, that is, perform the writing task). Changes in the SAT over the past decade reflect these debates about measurement and evaluation: "The 'new' SAT, introduced in 1993/94, reflects technical and philosophical changes as well as what students experience in today's classrooms" (Stewart and Johanek, 1996, p. 267). The SAT I is the norm-referenced measurement of reasoning skills, essentially a multiple choice examination testing critical thinking skills. The SAT II is a subject matter test, a test of subject knowledge, and includes a test of writing where the student actually writes an essay. The ACT examination is basically a combination of SAT I and II. Now comes word that one of our most influential psychologists, Robert Sternberg (2004), is working on the equivalent of the SAT III, an instrument to test intelligence more broadly, both in subject and skill, account for "creativity," and, eventually, influence college admissions criteria (Simon, 2004; Sternberg, 2004).

Throughout the century, curricular and evaluation debates have swung between two opposing positions: one knowledge-based, the other performance-based. Knowledge-based testing proponents argue, in the extreme, for a privileging of the subject or content, and are often trying to "conserve" the hegemonic (one) culture. Performance-based proponents argue, in the extreme, for privileging of the student's experience as a learner, and are often associated with a progressive/liberal ideology. In each case, assessment apparatus is shaped by profound differences in emphasis and value.

Where does this leave us, the teacher? When large-scale assessments are being used to predetermine not only what teachers teach, but also how they are to measure student learning and therefore, the forms in which students are taught, how can we attend to the day-to-day evaluation of student work? More disturbing, most of the current assessment policies and models actually remove the teacher and teaching from the equation. "Curriculum and instruction" have been replaced by "curriculum and assessment" in educational discourse during the past 20 years. We do not think this is a coincidence. When one adds in the attempts by neoconservatives to deny the value of teaching expertise in the attacks on professional licensing (and the support of various children's crusade programs like Teach for America), our profession has been diminished. Teachers are not asked to set goals, but to meet "benchmarks"—teachers are asked to teach to more "rigorous standards," not to help students become learners. Assessment is driving the education goals of the teachers and the local needs of the students. Wherever we look, teachers are under attack and denied agency (Apple, 1991).

But teachers deal with real students, not abstractions, and teaching is not adding information to a memory machine (E. D. Hirsch (1986) equates the brain to a computer hard drive). And though there are standardized tests, we are not trying to produce standardized students. But, in the end, much of school assessment mimics standardized assessment in its form and purpose. And, assessment for teachers is not merely about the attainment of some final state of education, but needs to account for evaluation that assists the teacher in making better judgments about teaching individuals.

In order to move from the hegemony of standardization, we as teachers need to move from "standardized" thinking to "situated" thinking. This change should even improve scores on standardized tests if done judiciously, because test taking is a situated activity. Most of the new knowledge about cognition (thinking) suggests that knowledge is situational. One knows things in relation to the enacting of this knowledge within particular situations or contexts. In other words, knowledge is as knowledge does. You can see the performance-based assumption here. In fact, in some instances, our tacit knowledge, the knowledge we act upon even if we cannot articulate it, is the most functional. Those of us, for example, who can ride a bicycle, would be hard pressed to explain how it happens.

Situated learning tells us that we do not learn *by* doing, but learn *in* doing (Lave and Wenger, 1991). Therefore, evaluation cannot be a measure of the result of learning but must be part and parcel of learning. Evaluation that is situated is an evaluation that not only represents the student's knowledge but also provides opportunities for the student to learn in performing the activity. Knowledge learned in doing suggests that learning is not the thing we do until the test, but what we do even while we are taking the test. Evaluation is not simply test taking but a process firmly embedded into the reflective teaching that we do in our daily professional lives.

As teachers, we need to work to change the conversation. The current content-based testing teaches students that knowledge is trivial, a sort of high stakes *Jeopardy!* (we choose this game and word deliberately). They do not see themselves as learners who engage in meaningful work with some sensible outcome, but as machine hard drives who will be punished (or suffer "consequences"—a key word in neoconservative discourse) if they do not memorize the appropriate information. We need to put the complexities of student learning at the center and to insist that students be engaged in intellectual activity—as participants in communities of practice—that calls upon their knowledge *and* experience. Our classrooms should provide adult teaching and direction while asking students to produce something of value to them as well as their peers, the school, and the community. Then evaluation—not just testing or assessment—flows in and out of the daily life of our work.

References

Apple, Michael (1991). *Ideology and Curriculum*, 2nd edn. New York: Routledge.
Applebee, Arthur (1974). *Tradition and Reform in the Teaching of English*. Urbana, IL: National Council of Teachers of English.
Gardner, Howard (1993). *Frames of Mind: The Theory of Multiple Intelligences*, 10th edn. New York: Basic Books.
Herrnstein, Richard and Charles Murray (1994). *The Bell Curve: Intelligence and Class Structure in American Life*. New York: The Free Press.
Hirsch, E. D., Jr. (1986). *Cultural Literacy: What Every American Needs to Know*. Boston: Houghton Mifflin.
Kincheloe, Joe, Shirley Steinberg, and Aaron Gresson (1995). *Measured Lies; The Bell Curve Examined*. New York: St. Martin's Press.
Kliebard, Herbert (1995). *The Struggle for the American Curriculum: 1893–1958*, 2nd edn. New York: Routledge.
Lave, Jean and Etienne Wenger (1991). *Situated Cognition: Legitimate Peripheral Participation*. Cambridge, UK: Cambridge University Press.
Lorber, Michael (1996). *Objectives, Methods, and Evaluation for Secondary Teaching*, 4th edn. Needham Heights, MA: Allyn & Bacon.
Ravitch, Diane and Chester Finn, Jr. (1988). *What Do Our 17-Year-Olds Know?: A Report on the First National Assessment of History and Literature*. New York: Harper & Row/Perennial Library.
Romer, Roy and Joy Fitzgerald (1996). A Vision for the Role of New Assessments in Standards-Based Reform. In Joan Boykoff, Baron and Dennie Palmer Wolf (eds.), *Performance-Based Student Assessment: Challenges and Possibilities*. Ninety-Fifth Yearbook of the National Society for the Study of Education, Part I. Chicago: University of Chicago Press.
Simon, Cecilia Capuzzi (2004). The SAT III? *New York Times*, Education Life (Section 4A), p. 15.
Sternberg, Robert (2004). *Wisdom, Intelligence, and Creativity Synthesized*. Cambridge, UK: Cambridge University Press.
Stewart, Donald M. and Michael Johanek (1996). The Evolution of College Entrance Examinations. In Joan Boykoff, Baron and Dennie Palmer Wolf (eds.), *Performance-Based Student Assessment: Challenges and Possibilities*. Ninety-Fifth Yearbook of the National Society for the Study of Education, Part I. Chicago: University of Chicago Press.

Chapter 5

The Cult of Prescription—Or, A Student Ain't No Slobbering Dog

P. L. Thomas

This is a true story, but it is one most people never see—a story that reveals what is behind the things we do in our schools:

> I recently sat talking about the teaching of writing with a principal I have known for nearly thirty years; the principal is one of the most dedicated educators you can find, and she was an outstanding teacher for many years before moving to administration. I was explaining to her the inherent problems with teaching writing through prescription—forcing students to write from templates similar to the traditional five-paragraph essay. Her response? The principal replied something like this—"Why, Paul, if students can't even do exactly what they are *told* to do, how are those students going to write papers on their own?"

Everything that is wrong with our public schools, everything that keeps our public schools from fulfilling its role in a free society—as Thomas Jefferson envisioned—is captured in this scene. This principal personifies the cult of prescription that lurks just under the surface of our schools; ironically, our free society trains students in America to *do as they are told* before they somehow earn the right to make their own course through this life.

Currently, our public schools are under attack; critics from politicians to radio pundits to newspaper journalists are telling us what is wrong with our schools—and how to improve them. These criticisms and silver-bullet solutions soon seep into the daily thoughts and beliefs of all of us. Yet, most who have these discussions about education simply do not know about or fully understand the many assumptions and agendas that drive this thing we call public schools.

Are our public schools failing us in some way? Yes. Do our schools need reform? Yes. But the problem we are facing is that the great majority of criticisms and solutions are resting squarely on false assumptions and noneducational agendas that make both the criticisms and the solutions misleading and harmful. Here is one thing you will almost never hear: A troubling dynamic has haunted fruitful school reform for decades—those without the training or the expertise to reform education manipulate the misconceptions and inaccurate perceptions of the average citizen for political or financial gain. I attempt to uncover in this chapter those assumptions and agendas that most of us never see—even though criticisms and discussions of education remain on our lips almost daily.

I have written elsewhere about the Frankenstein nature of the American psyche (Thomas, 2004)—Americans have a near obsessive love for and complete misunderstanding of numbers and all things scientific—and I have warned before about the manipulation of politicians who are more and more using schools, standards, and high stakes testing for political gain (Thomas, 1999). The blunt truth is that most people who speak about public schooling are victims of false assumptions concerning behavioral learning theory, capitalistic ideology, and testing. When the assumptions are misguided, the criticisms and subsequent solutions are misguided as well. Although public education is possibly the greatest commonality among the American people, most of us simply do not know much about the theories and beliefs that dictate how our schools function on a daily basis.

Just because Pavlov was successful in conditioning a dog to salivate over a bell instead of food does not mean that our students should be treated like canines in our big experiment called school. You see, *a student ain't a slobbering dog*. Let us first look at the false assumptions, those things that for the most part are simply never clearly stated when we debate education—where the assumptions come from and how they impact our schools. Then we can begin to have a frank and honest discussion of where our schools are failing and how we might begin to create the kind of schools our students and our free society deserve. It turns out that when we are talking about schools, what we do not know *will* hurt us.

The Cult of Prescription

Paulo Freire (1993) makes this charge:

> One of the basic elements of the relationship between oppressor and oppressed is *prescription*. Every prescription represents the imposition of one individual's choice upon another, transforming the consciousness of the person prescribed to into one that conforms with the prescriber's consciousness. (p. 28; italic in original)

Freire's broad and philosophically grounded complaint is not removed at all from other concerns voiced now and more directly on the state of education and educational reform in America; Osborn and Gayle (2004) make this observation about the most recent educational reform mandated by the federal government: "The No Child Left Behind Act, for instance, encourages rote learning by aligning highly specified lessons with mechanized tests." What many fail to recognize is that most educational reform initiatives are prescriptions, and for this very reason, we should reject them.

From politicians and pundits to administrators and classroom practitioners, our schools are steeped heavily in the cult of prescription—a way of thinking that grows from our embracing behavioral learning theory, our capitalistic zeal, and our mindless allegiance to standardized testing. Each of these must be exposed and explored closely, and then abandoned if our schools are to prepare young Americans to become a contributing part of a free and democratic society. If our goal is otherwise, or if we are content to allow education to happen uncritically *to* our children, then we should keep along this traditional imposing course.

Behavioral Assumptions

In Barbara Kingsolver's *Prodigal Summer*, the main character recalls her father telling her, "*There's always a reason for what people say, but usually it's not the reason they think*" (2000, p. 362; emphasis in original). I would extend this idea to what people *do* as well.

Americans tend to be a pragmatic people; we scoff at theory, we scoff at those things merely academic. Yet, whether we are conscious of the fact or not, all things we say and do are grounded in theory. We say and do things that reflect what we *believe* to be true, whether it is true or not, whether we can express those beliefs or not.

This is disturbingly true in public schools. The reasons behind what we do, say, and think in education are primarily driven by our allegiance to behavioral psychology—assumptions about human nature and the acts of teaching and learning. The profound impact that behavioral assumptions have on teaching and assessment in our schools—along with its impact on the ways in which we manage the classroom, including discipline procedures and policies—directly conflicts with our expressed reason for public schooling: To support a free democracy. When classroom practices steeped in behavioral assumptions are exposed, we see that those behaviors are both morally and educationally suspect.

Brooks and Brooks (1999) identify the essential contrasts between the traditional classroom grounded in behaviorism and the progressive classroom grounded in constructivist learning theory. What many simply do not know is that the vast majority of teachers have been trained as if behavioral psychology is not only the best explanation for teaching and learning, but also the *only* valid theoretical grounding for educating children. However, at least one well-supported and legitimate alternative to behaviorism exists: constructivism. This alternative proves to be better suited to public schooling in a free democracy, but as we will see, those who control public schools have many different goals, least of which is supporting our democracy. When exposed for what it is, the traditional classroom guided by behavioristic principles functions under a few intellectually paralyzing assumptions, ones that contribute to many things other than individual empowerment and democratic principles:

1. Material to be learned is constant and easily identified, thus irrefutable.
2. Learning comes from teaching that is imposed from without. Teachers dispense knowledge into students as if those students are simply receptacles.
3. Human nature *requires* that learning and behavior be controlled by a system of rewards and punishments (stimulus–response); good classwork deserves high *grades*, and poor behavior deserves swift and strict *punishment*.
4. The content of learning is best learned in small pieces and in a linear fashion. Parts are more easily learned than wholes, especially if they are sequenced.
5. Being wrong or making mistakes must be *avoided* by learners. Proper rewards and punishments can help teach learners to avoid being wrong or making mistakes.
6. The learner is essentially a blank disk drive waiting to have information saved in neat files and folders by a knowledgeable and benevolent (adult) computer programmer.

The list of highly esteemed behavioral psychologists is long, and the history of our support for their theories is rich. How then can we refute the value of such assumptions in our schooling of children? Easily.

First, let us look at *why* behavioral psychology has such a power over our schools. Though this appears not to be common knowledge, behaviorally grounded practices in schools dominate because they are easily standardized, easily managed, and easily measured—*not* because they are morally superior, *not* because they are sensitive to the humanity of children, *not* because they support the needs of a free democratic people. In fact, teaching and learning driven by behavioral assumptions are morally questionable, prone to dehumanizing effects, and contradictory to democracy and individual freedom. Let us look at the assumptions of behaviorism from above and see why they are counter to the potential our schools could offer a free people:

(1) The content of learning is easily identified and essentially fixed. This assumption is the most telling of the tendency of behavioral assumptions to foster indoctrination, not learning. The truth about the content of learning is that knowledge is ever-shifting and—as is the nature of intellectual discourse—*debatable*. In fact, it is the nature of learning to debate ideas—not to memorize information simply to regurgitate it later for the approval of some authority.

(2) "To teach" means to dispense information to learners. Possibly the most damaging assumption of behaviorally based education is the idea that the teacher is the sole source of what one should know, that the teacher's job is to distribute that knowledge that he embodies. All of us know that the things we have learned most dearly are things that we somehow embraced as active learners; what people have merely told us has generally rolled off our backs like so much water. The effective teacher is much more like a supportive coach than a judgmental preacher.

(3) Human nature requires that learning and behavior be controlled by a system of rewards and punishments (stimulus–response). As Kohn (1996) asserts, behavioristically grounded classrooms are driven by a very dark view of human nature; the assumption is that all humans must have extrinsic rewards and punishments or nothing can be accomplished—or worse yet, there will be chaos! As I discuss later, I believe most of children's need for rewards has been *learned* in a society that has sold its soul to capitalism. Learning, striving, struggling, and living are all both the act itself and the reward simultaneously. And humans by nature are actually eager learners and active participants in life. School—with its relentless system of punishments and rewards—drains the life out of learning and out of children to the point that children *seem* able to function only within a system of rewards and punishments. Schools actually create a self-fulfilling prophesy. The sad irony is that the behavioristic assumptions of school practices have conditioned students to be that which behaviorism claims they are!

(4) Parts are more easily learned than wholes, especially if they are sequenced. It seems so obvious that this assumption *has* to be true that I am always nervous to refute it. But I do refute it. This aspect of learning is counterintuitive. Brain research is confirming over and

over again that *most* people (some say as much as 80 percent) actually learn from whole to part. More students learn best dealing with the big picture first; they actually grasp the essential concept before they are able to explain the process, before they are able to analyze the concept. The *rare* person is analytic by nature. Brain research is also showing that we are quite chaotic learners as well. It does *seem* likely that learning things in an organized sequence would be what is best for initial learning, but, again, this appears not to be true. Schools function under the assumption that *all* students are analytic and linear learners. In fact, most are global learners who grasp ideas and information in a chaotic manner. Analytic and linear teaching is *more manageable*, not better supported by research, not better suited to support our democracy (though it has some benefit to creating more pliable *workers*).

(5) Mistakes and errors should be avoided by learners. From our red pens and our verbal reprimands, we teachers teach children daily that good students should not make mistakes. Constructivistic learning theory, however, embraces the *value* of error in learning (Brooks and Brooks, 1999). Mistakes are actually a first step toward rich understanding. As well, risk taking is a key to deep learning. If children are paralyzed by the fear of doing wrong (and of the subsequent punishment—thus supplanting any concern by the learner for that act that has been deemed "wrong"), they are highly unlikely to take risks; as a result, they are unlikely to come to understand anything in a real and deep way.

(6) The learning brain is a blank slate. Again, as brain research expands, we know that this is a deeply flawed assumption. As one example, linguists are fairly certain that even language and grammatical competence are pre-wired in the brain (Pinker, 1994). All new learning is in some way measured against or merged with prior knowledge when students learn. It is a great mistake to believe that learning occurs in a vacuum.

The behavioristically grounded classroom, the traditional classroom, values a passive learner and embraces indoctrination and rote memorization as valuable and even desirable learning. As I discuss further in the last section of this chapter, the passive learner is the least desirable student for our classrooms and ultimately our society. Behavioristic assumptions create an incredibly low expectation for children—they must be trained—and determines the lowest path for most students in their *living*. The passive learner is no learner at all, having had her humanity stripped from her by a series of punishments and rewards over more than a decade of her most impressionable years on this planet.

Capitalistic Ideology

While our schools are essentially behavioristic psychology in practice, our larger American society is primarily capitalism in action. It is difficult, I admit, to clearly separate the impact behaviorism and capitalism have on how Americans do *everything*, from advertising our products to training our dogs to educating our children. Actually, I believe that our bigger commitment to capitalism contributes to our somewhat mindless allegiance to behaviorism. It is not a far leap from stimulus–response in the classroom to the 40-hr workweek that ends with a pay check.

My criticisms of our public school system grow from my belief that our schools are failing our *democracy* and the *humanity* of all children. Most criticism, most popular and political criticism springs from a belief that our schools are failing our businesses, and our economy. Those are drastically differing charges. Engel (2000) makes a Deweyan claim that Americans value capitalism more than democracy—and nowhere is it shown more vividly than in our schools. He believes, "A democratic school is one that . . . tries to enable people to create their own world collectively rather than fit into one that is created for them" (p. 65). But our schools practice the prescriptive tendencies of behaviorism and bend to the political and business forces that charge schools with reinforcing the economic interests of our society; we need workers who will fit the corporate mold, they argue, or America is doomed! We are not being told that most criticisms and reforms of schools are driven by political and business interests—agendas that grow from how schools can best support someone's wealth or someone's power base. Is the call by some to end public schooling and open education to the free market a move to improve learning or to create a new market where money can be made?

Combined with the behavioristic assumptions, capitalism fails our schools and our students in these ways:

1. Faith in free market ideology—the competition model—as a panacea for any social ill casts a misguided cloud over our public school system, which *appears* to be a monopoly, but only in a simplistic view of monopolies. Political and business criticisms of schools often spring from the claim that public schools can never work since they are monopolies and since monopolies are anathema in a capitalistic society. Yes, in the business world, a monopoly fails capitalism. But schools are not providing a sole service *for inflated financial gain or for the unfair concentration of power*; in other words, public schools are *not* a monopoly (no more so than the military, the police force, or the judicial system, that is). Simply being the only source of a service is not a monopoly; being the sole source for inflated financial gain or

for unfair power is or there is no harm in being the sole provider. Even in a capitalistic society, a few social entities must exist for the ultimate good of the entire society; public schools is one such entity.

2. Capitalism and the American Dream written as *becoming rich* reinforce two weaknesses in our society—the valuing of extrinsic rewards and the inflated trust we place in numbers. That things can be quantified and are quantified means a great deal to Americans. The business world learned quickly that products sell if they are *scientific*, if there are numbers to support why a product is the best. If something is valued, it is quantified, is what we Americans believe. The bigger the number, the more it matters. Teachers make relatively little money so they do not really matter—especially when compared with professional athletes or movie stars. Worse still, public education is often called "free" (though it certainly is not); thus, how can anything free be of any value?

3. Capitalism suggests that ends matter more than means; for many Americans, whether in school or at work, what comes at the end justifies, or at least overshadows getting to that end, the means. In school, the grade comes to be more valuable than the learn*ing*; at work, the paycheck matters more than the work*ing*. Learning, the kind that empowers each individual and enriches any community, is a journey and not a destination. Capitalistic zeal for the finish line poisons real learning, and destroys any appreciation for the journey.

That we are a people more deeply committed to capitalism and our own ideologies than to democracy and free society is tremendously significant in the way we run our schools. "We have factory-based schools in an Information Age—and no factories. . . . This factory-based approach, however, is locked in by political gridlock . . . [which] encourages rote learning by aligning highly specified lessons with mechanized tests," observes Osborne and Gayle (2004).

In educational jargon, schools are referred to as "plants." This is no simple metaphor. The business paradigm has been neatly placed over the behavioristic skeleton of our schools to create a powerful being few people face or question.

Standardized High Stakes Testing

The assumptions of behaviorism and capitalism are both emboldened by and productive of numbers; we love numbers in the United States. For our schools, the numbers are increasing exponentially as the standards movement and the high stakes testing bonanza have mushroomed, fertilized as they have been by the fecund criticisms heaped on our public schools by politicians, pundits, and business leaders during the last two decades.

The only thing that can match America's trust in behaviorism as how we should teach children is our unwavering faith in standardized testing. From IQ to the SAT to GPA, Americans rely on tests and the numbers from these tests to label our students as smart—or not.

Here is what few people know: The big tests of our society—the SAT, for example, which is designed to predict college success but is only the third best indicator of this success, behind GPA and courses taken, both of which cost the student nothing (Thomas, 1999)—do not prove what we think they prove, and they are draining our schools of valuable time and money. Popham (2003), himself an expert in educational assessment and high stakes testing, stands as one of many who are quick to warn that standardized testing is not all that most believe it to be. Often, mass produced and scored tests do not adequately measure the things that matter, but like behaviorism, certain types of machine-scorable testing have an upper hand in education because it is easier to manage and cheaper to implement—neither of which has much to do with being educationally sound or productive. As well, testing has the added veneer of being easily quantified, thus feeding the American belief in numbers as truth.

The computer boom has helped to exacerbate the high stakes testing movement. Many states and even the SAT are moving toward not only scoring multiple-choice tests by computer, but also grading student compositions by computer. These moves have and will continue to increase computer-aided instruction and assessment in classrooms and for high stakes testing. Both the use of computers and the trust we place in tests will simply increase all that is wrong with schools—paralyzed as they are with the cult of prescription.

Standardized, high stakes tests fail students and society for many, many reason, but let me offer a few simple but relevant examples of how they fail us:

1. Many, if not all, universities offer degrees in areas such as art and music. What have the type of verbal and mathematical questions on the SAT got to do with ceramics or playing the cello? Nothing. Why then must an aspiring artist make a certain score to enter college to major in ceramics and sculpture? It makes no sense.
2. How does our society accept the settling of the best high school or professional football teams? State championship *games* and the Super Bowl, right? Why do these events—mere entertainment—not settle matters with multiple-choice tests than we can easily run through a machine for a score? We have computers involved in the determination of college football's champion, and no one is happy about that! But we are somehow content to have a computer grade a student *essay* and then allow that score to determine whether that student can graduate high school or attend college?

To be honest, the flaws with the current state of high stakes testing are far more severe than my simple examples; better people than I am are beginning to make that case. Here, I simply want to state flatly that our blind trust in testing in the traditional sense is harming our schools, our students, and our society, *but from the loudest critics of our schools you will hear just the opposite*. For them, we need higher standards and more testing—not just of students but of teachers too. The current mentality on testing is captured well in an apt Southern expression: *Weighing the pig won't make it fatter*.

New Paradigms, New Schools, New Honesty

Several years ago, I called for a new honesty in education (Thomas, 1999), but only recently have I come to realize what I *really* meant by that call. You see, the most fervent behaviorists, capitalists, and positivists—those who *already* control our schools—have been leading an assault on public education because the schools are not entrenched *enough* in these areas. Those of us who are educators because we love the humanity of children and the righteousness of individual freedom and democratic values have unwittingly coalesced to defend the current system, though it is in contrast to all that we hold sacred.

Recently, I witnessed first-hand how pervasive the cult of prescription is *even within the educational establishment*. One of the hot ideas and subsequent books and materials of the year is the concept of backward design, embodied by the work of Wiggins and McTighe (1998), *Understanding by Design*. To me, this concept of backward design is simply prescription light, and it indirectly reinforces all of the problems I have detailed above. This is difficult to state bluntly, but few have looked closely enough at the concept of backward design. *Determining for a learner what is to be learned or performed and then offering that learner a template to fulfill those expectations is a highly manageable and efficient process but it is not education; it is training*. Educators—mainly administrators—have fallen prey to trying to outdo the critics, without considering the flaws in the political and commercial calls for school reform.

Our schools need neither the current political criticisms and solutions nor the defense of practitioners; schools especially do not need the materials and programs to fulfill those political demands. Our schools need a new and honest unveiling of those things that are wrong and how we can steer this educational ship onto a course that is true and good. The truth is, public schools in America do many wonderful things, but fail in some truly serious ways as well—but not the ways in which George W. Bush, fundamentalist Christians, the business lobby, or textbook publishers claim.

Let us leave this discussion with a serious skepticism for the cult of prescription that permeates our schools and with words from two educators who express honestly the schooling that could be, the schooling that should be: "Education will unfit anyone to be a slave. ... Education tears down walls; training is all barbed wire" (Ayers, 2001, p. 132), and "At his or her best, the hungry student is the constructively skeptical student" (Sizer, 1992, p. 54).

Students should not be slaves to rewards and punishments, should not be slaves to a pay check, and should not be slaves to the score on a high stakes test. Teachers and students should not be slaves to materials, textbooks, worksheets, and computer programs that prepare students for *the test*. That is all barbed wire. The promise of a free democracy lies in the hands of young minds, hungry and skeptical— free of schooling that is mere indoctrination.

References

Ayers, W. (2001). *To Teach: The Journey of a Teacher*, 2nd edn. New York: Teachers College Press.
Brooks, J. G. and M. G. Brooks (1999). *In Search of Understanding: The Case for Constructivist Classrooms*. Alexandria, VA: Association for Supervision and Curriculum Development.
Engel, M. (2000). *The Struggle for Control of Public Education: Market Ideology vs. Democratic Values*. Philadelphia: Temple University Press.
Freire, P. (1993). *Pedagogy of the Oppressed*. New York: Continuum.
Kingsolver, B. (2000). *Prodigal Summer*. New York: Perennial.
Kohn, A. (1996). *Beyond Discipline: From Compliance to Community*. Alexandria, VA: Association for Supervision and Curriculum Development.
Osborn, H. and M. Gayle (June 1, 2004). Let's Get Rid of Learning Factories. *Los Angeles Time*. Retrieved June 4, 2004, from: http://www.charlotte.com.
Pinker, S. (1994). *The Language Instinct*. New York: Harper Perennial.
Popham, W. J. (2003). *Test Better, Teach Better: The Instructional Role of Assessment*. Alexandria, VA: Association for Supervision and Curriculum Development.
Sizer, T. R. (1992). *Horace's Compromise: The Dilemma of the American High School*. New York: Mariner Books.
Thomas, P. L. (1999). A New Honesty in Education—Positivist Measures in a Post-Modern World. *Contemporary Education*, 71, 1, 51–55.
Thomas, P. L. (2004). *Numbers Games: Measuring and Mandating American Education*. New York: Peter Lang.
Wiggins, G. and J. McTighe (1998). *Understanding by Design*. Alexandria, VA: Association for Supervision and Curriculum Development.

Chapter 6

School Leaders, Marketers, Spin Doctors, or Military Recruiters?: Educational Administration in the New Economy

Gary L. Anderson

The recent No Child Left Behind (NCLB) legislation is the culmination of a series of commissioned reports and policy initiatives that have created a new policy environment for principals and superintendents. Major components of these policies include high stakes testing, increased competition for students, and recruitment of students for the military. Never before have policies so consistently bypassed the professional judgment of administrators and teachers. This is not by accident. A mixture of business models and free-market ideology have descended on education, threatening the very ideas of leadership and public education.

Sergiovanni (2000) describes the current dilemma for administrators as an encroachment of the systemworld on the lifeworld of the school. The *systemworld* consists of the management designs, policies, rules, and schedules that provide a framework for students and teachers to engage in teaching and learning. The *lifeworld* is the culture of the school and is represented by the values, norms, and beliefs that determine the social interactions among students, teachers, and administrators. Both the lifeworld and the systemworld are essential to the school organization. When the two are in balance, they function symbiotically to create schools and classrooms in which system efficiency is in harmony with the lifeworld where teaching and learning takes place. However, in many schools, the systemworld dominates the lifeworld. A dominant systemworld destroys the fabric of the school culture creating isolation, alienation, and a loss of a sense of professionalism.

This describes the current standards movement in education. Because in the past teachers could close the classroom door and ignore reforms that failed to understand their local context and because administrators had considerable autonomy, the lifeworld of schools was often buffered from the constant demands of reforms that came and went. In the new context of high stakes testing, instructional "coaches," administrative "walk throughs," and school report cards published in local papers, the systemworld has begun to "colonize" the lifeworld of schools. In a system that forces the classroom door open and is obsessed with high standardized test scores, leadership becomes so circumscribed that it can be exercised by neither administrators nor teachers. In this way, the new reform movement is essentially replacing leadership with standards and standardization.

So the decline of leadership is the decline of the lifeworld of the school in the face of a resurgence of Taylorist forms of bureaucracy, scientific management, standardization, social engineering, and competition that steers the system by exerting pressure from the top, while allowing limited local "autonomy" over (mostly trivial) decisions at the bottom. Thus, site-based management becomes a form of decision making about the preferred *means* to achieve the *ends* determined elsewhere. Currently, in many school districts, such as Los Angeles Unified, where the Open Court Reading Program is required in all schools, even the instructional *means* are taken out of the hands of the educators. Principals are left with managing their school culture and, as I discuss in more detail below, managing the image of their schools or, in the case of superintendents, their school districts.

However, stripping teachers and principals of professional autonomy is taking place through a moral discourse of "Leave no child behind." In this sense, teachers, principals, and superintendents, who are now referred to as "the educational establishment," can no longer be entrusted professionally with the education of America's youth. Thus, the systemworld exerts its considerable pressure on the lifeworld, shaping a more fearful and docile school culture in which school professionals tow the line or take the increasingly popular options of early retirement or leaving the profession altogether.

Nevertheless, while most educators find this tendency troubling, it is true that schools in the past have not done a very good job of empowering students and their communities, and administrators and teachers have often been implicated in practices that keep schools tracked by socioeconomics and race and closed off from the surrounding communities, particularly in low-income areas. Therefore, from an equity perspective, providing greater autonomy for administrators and teachers is no guarantee that they will "do the right thing" or even know what "the right thing" to do is.

One thing is certain: The constant pressure on administrators is causing many to engage in extreme forms of impression management. Engaging in public relations has always been part of a school administrator's job, and in this age of public school bashing, highlighting the positive things about one's school is a necessary counter to the sensationalist news that frames minority youth as criminals and urban schools as on the verge of collapse.

There is also nothing wrong with pressuring administrators to make sure that all students in their schools are getting the best education that can be provided with the resources these schools are allocated. Good administrators have always had a low tolerance for low expectations and mediocre teaching. However, this new reform, led primarily by business leaders and politicians, has upped the ante, leading to the creation of what Murray Edelman calls the creation of a "political spectacle." In his book *Constructing the Political Spectacle*, Edelman (1988) argued that elites constructed the political spectacle through their manipulation of language and media, the evocation of enemies, the use of rational language to hide political interests, the creation of a sense of crisis, and the conversion of active citizens to passive spectators. This has led to what Berliner and Biddle (1995) have called "a manufactured crises" in education that is attempting to undermine the public's faith in public education. In many ways, administrators are taking their lines from a national script that has turned traditional notions of public relations into a new ethos that allows for the outright manipulation of information with the goal of deceiving the public.

Public Relations on Steroids: Deceiving the Public

At the national level, Edelman's political spectacle has become so obvious as to be self-evident. Before the U.S. entry into the First Gulf War in 1991, a tearful 15-year-old Kuwaiti girl named Nayirah testified before Congress, with news cameras rolling, that she had witnessed Iraqi soldiers stealing babies out of incubators in a Kuwait city hospital. Later it was revealed that the girl was the daughter of a Kuwaiti diplomat in Washington, and that the Kuwaiti government had hired the American public relations firm of Hill and Knowlton to stage the testimony in Congress (MacArthur and Bagdikian, 1993; Kelly, 2002). This vilification of Iraqis signaled the construction of a new enemy and was a turning point in the U.S. public's support for the war. In the more recent conflict in Iraq, a dramatic rescue of a courageous young American female soldier, Jessica Lynch, who had allegedly been captured in a blaze of gunfire and later mistreated by her Iraqi captors was miraculously captured on film. Back in the United States, this "war hero" immediately became front-page news and the subject of a television movie. It later turned out that her injuries were due to a vehicle accident, that an Iraqi medical team had nursed her back to health and had tried to deliver her to the U.S. troops. Although some facts are still in dispute, the U.S. troops apparently fired on the vehicle, which returned to the hospital so the rescue could be staged (Neuman, 2003). The notion of a "theater of war" has taken on a more literal meaning in this age of political spectacle.

While the current school reform spectacle is perhaps less dramatic, it is becoming equally evident. Rod Paige, the secretary of education who was brought to Washington by George Bush, was the superintendent of the Houston Independent School district from 1994 to 2001. This district had been touted as a jewel in the crown of the "Texas miracle" in school reform circles and in 2002 won a 1 million dollar prize as best urban school district in the country from the Los Angeles–based Broad foundation. An article on the front page of the *New York Times*, July 11, 2003 reported that "the results of a state audit found that more than half of the 5,500 students who left their schools in the 2000–2001 school year should have been declared dropouts but were not. That year, Houston schools reported that only 1.5 percent of its students had dropped out" (Schemo, 2003, p. 1). The audit recommended lowering the ranking of 14 of the 16 audited schools from the best to the worst.

Just as teachers know how to "perform" for evaluators, administrators know how to "perform" for politicians and business leaders who have crafted an accountability system that is mostly stick and very little carrot. Punishment, humiliation, and loss of jobs await those administrators whose schools fail to "perform" up to standards. The recent Annual Yearly Progress requirement of NCLB is pitched at such an unrealistic level that it is likely to provide even more incentive for administrators to manipulate data. This mixture of pressure from above and semiautonomous "learning communities" or "teams" is a model borrowed from the corporate world.

The Corporate Influence on Educational Administration

Much of the new language that has entered the lexicon of educational administration came from an army of workshop leaders and professors who taught the principals of total quality management (TQM), which became popular in the 1980s and 1990s. In reality, TQM had already been critiqued within the corporate world, when educators picked it up. Terms like "continuous improvement," "teaming," "customer," "quality," and the like became in this period part of the vocabulary of administrators. TQM principles were even promoted in classrooms through the Baldridge approach to instruction.

TQM devolved decision making to workers, promoting teaming and site-based decision making. However, as Harley (1999) and many others have pointed out, "while strategic decisions continue to rest with management, there is a devolution of responsibility for tactics to the core workers" (p. 316). This essentially becomes a new more sophisticated motivation theory in which workers—or teachers—have the illusion of control over their workplace, while real control is consolidated at higher levels of the system. Thus teachers, according to this business model, are encouraged to "take ownership" and "buy into" someone else's agenda. In the corporate world, workers are becoming increasingly aware that while they are being "empowered" on the shop floor through participation in selected work-related decisions, their unions are being busted, their companies downsized, their jobs moved overseas, and their salaries and benefits slashed. A similar trend can be seen in education where funding is slashed and top-down testing regimes coexist with teacher "empowerment" and "autonomy" through group decision making and teams. Edelman (1978) put it more bluntly,

> Participation in group meetings has often been obligatory: in China, in Russia, and in Nazi Germany, just as it usually is in mental hospitals, in prisons, and in high schools that emphasize student self-government; for it helps evoke popular acquiescence in rules that would be resisted if authorities imposed them by fiat. (p. 121)

The other aspect of TQM that has infused educational administration is the notion of statistical control and the elimination of variance. In most businesses it makes sense to want to produce products that eliminate variance. A quality product is one that does not vary, such as a McDonald's hamburger or an air filter for an automobile. Any variation is viewed as a defect and the use of statistical control helps to eliminate variation. This notion has essentially been lifted from business and applied wholesale to education through current testing regimes that exert statistical control over student achievement. The problem is that the core technologies of business and education are fundamentally different. Successful student achievement depends on addressing variation in students and the essence of education is helping students find their own individual and unique self-actualization.

Corporate influence in education, while not new, is stronger today than at any time in our history. Goals of schooling, such as building democratic citizens or providing opportunity, have been replaced by a concern for forming human capital for a global economy. Corporations have given us much that is good. Business as a social enterprise is a cherished foundation of our economy and society. Few economists today would argue that any country can prosper without a vibrant private sector. However, humans are not merely *homo economicus* and our schools were never meant to merely serve the needs of business. Corporations have a place in our society, but sovereignty belongs to the public. Without a government and public spaces independent of corporate control, the notion of "public" in public schooling is called into question.

Testing the Testers: An Exam for Administrators

After years of testing teachers and students, school leaders are now the targets of the testing industry. Educational Testing Service (ETS) has developed an examination based on the Interstate School Leaders Licensure Consortium (ISLLC) national standards for school administrators, which are replacing state standards across the United States. The ETS exam is required for school administrator certification in several states. These new standards and the exam that enforces them are driving the preparation of future school administrators.

While the new standards themselves are not that different from the previous ones, the ETS exam is another story. The Registration Bulletin for the School Leaders Licensure Assessment (1999) that is published by the ETS provides examples of test items and exemplary responses. I have documented elsewhere in more detail the ways this exam encourages a narrow public relations view of school administration (Anderson, 2001, 2002). The language of the standards themselves is a utilitarian language linked to business and the economy. The ISLLC standards contain a largely noneducation vocabulary with terms like "alignment," "strategic planning," "operational procedures," "core technology," "entrepreneurally," and "marketing strategies" predominating. How words and metaphors come to orient professional practice is a research agenda that is still in its infancy in education.

The exam reinforces a practice devoid of critical thought and is focused largely on smoothing over conflict and contradictions with public relations techniques. The following is an answer in the exam bulletin that was considered exemplary.

The broad based issues the school must resolve are in the areas of *communication and public relations,* . . .
There is a need for *communication and p.r* . . .
Whenever there is a letter writing campaign, this issue as a *public relations* concern must be addressed . . . or *a domino effect will likely occur.* (A single letter from a parent concerned about the school's use of cooperative learning was included in the sample exercise.)
The public at large also *needs to be educated.* Although the PTA is *an effective arm of the school,* there needs to be budgetary line items allotted to *parent training.* (p. 13)

Exemplary answers are generally defensive, reactive, and have a deficit view of low-income communities.

Learning to Put the Right "Spin" on Answers

The notion that professionals have espoused theories and theories-in-use that are seldom isomorphic is not news. This is a basic premise behind why many professional education programs teach reflective approaches to professional practice. However, as I have described elsewhere, a major role of administrators is to legitimate their organizations to multiple constituencies (Anderson, 1990). In the case of educational administrators, they must legitimate not only their own organization, but also an institution—public education—that is currently in crisis. This legitimation role, which requires different discourses of legitimation for different constituencies (central office, faculty, parents, community, media, etc.) produces a discourse similar to that of presidential candidates who use language in such a way as to not offend any particular constituency.

Thus, administration programs increasingly are in the business of providing future administrators with "safe" discourses that will not offend pluralist interest groups. However, as Schattschneider (1975) has pointed out, "the flaw in the pluralist heaven is that the heavenly chorus sings with a strong upper-class accent" (p. 35). Some constituencies within a pluralist political framework have more power than others, and expanding participation to wider groups, such as students and communities, threatens the force field of power that maintains a particular status quo and often leads to a greater recognition of interests that are in conflict (Schattschneider, 1975). Administrators are seldom rewarded for expanding the scope of participation, unless it can be done in such a way that it does not result in a significant shift in power relations. At the same time, the legitimacy of public schools requires discourses of democracy and equity, just as they require discourses that reflect scientific, research-based practices and reforms that can socially engineer increased student outcomes.

Given the nature of these demands on educational administrators, the gap between rhetoric and practice is viewed as necessary. With their fingers in the political wind, administrators are taught to seek the path of least resistance, rather than take risks. The ambiguities, ambivalences, and contradictions that run through the standards also run through the exemplary responses to the exam questions. Thus, it is acceptable to espouse an explicit public relations approach to community "buy in" for decisions largely made elsewhere, while simultaneously calling for community participation in decision making.

None of the answers throughout all of the sample questions in the bulletin, including those that were exemplary, showed any indication that any respondent had read any professional literature. Not a single author was alluded to, much less cited in the responses. Occasional references are made to specific instructional programs like Reading Recovery or general approaches like cooperative learning, but there is little indication that there are important conceptual debates over instructional methods, approaches to school governance, or the role of schools in society. There is no indication that any of these future educational leaders read anything beyond technical manuals and highly condensed administrative textbooks. One of the respondents lost points because the scorer felt that "throughout, responses are weakened by suggestions for solutions that are *outside the principal's control* (emphasis added), specifically the suggestions to redistrict and to increase the staff" (p. 48). Clearly, reading reports by the Children's Defense Fund on how the national budget neglects children would be a waste of time for aspiring principals, or seeing one's self as an advocate for policies outside the narrow confines of one's school would be overstepping the limits of one's role. The implications for certification and graduate degrees in educational administration, if these exams drive the curriculum, are depressing. The notion that well-educated individuals with a commitment to democratic values should lead our schools is replaced by the notion of cohorts of glib technocrats who have a brief, clear, and "convincing" answer to any problem.

From Public Relations to School–Community Alliances

Most research and writing in educational administration today adopt, either explicitly or implicitly, the image of the business CEO as the prototype of the effective educational leader. However, other prototypes exist. For instance, Martin Luther King drew on the social organization that existed among African Americans in churches and other community organizations where they congregated. Educational administrators have no problem with school–community "partnerships" defined as local businesses, but tend to shy away from partnerships with local communities. Crowson and Boyd (1999) are eloquent on this point:

> The need to preserve strong norms of professional discretion against private-regarding parents and narrow-minded communities was a theme as early as 1932, in the work of Willard Waller. Generations of school administrators in the U.S. have been trained around the dangers of losing managerial control to the "politics of their communities." (p. 11)

However, recent scholarship is documenting a growing alliance between school administrators and community organizing groups, such as ACORN, the Industrial Areas Foundation, and other Inter-Faith groups (Gold et al., forthcoming; Shirley, 1997, 2002). Administrators obviously cannot work outside the system as King did, but they can build alliances with communities.

This new scholarship on community organizing for school reform distinguishes what Shirley (1997) calls accomodationist forms of parental *involvement* from transformational forms of parent *engagement*. Most current approaches to school–community relations are school centered and accommodate parents through involvement rather than using them as a resource that can challenge and transform schools through authentic engagement that can lead to long-term improvement.

Unfortunately the current paradigm of school administration sees the surrounding community as a threat rather than as a resource and ally. Community organizing groups are aimed at building community capacity so that low-income communities can advocate for what they need. This can be a powerful network for school administrators to tap into. Organized communities that are linked to schools have helped schools get the attention of local political leaders to get access to additional resources and infrastructure improvements such as bond initiatives, after-school programs, more crossing guards, and improved traffic patterns in school areas. Once administrators realize that local communities have concerns that overlap with theirs, they are more willing to take the necessary risks that democratic participation always entails, such as increased conflict and some loss of power. Schools with intimate connections to local communities are also in a better position to build on community "funds of knowledge" (Moll et al., 1992).

In their study of 19 community organizing groups supporting school reform Gold et al. (2004) found that authentic community involvement in the leadership of the school helped to sustain positive changes in the face of administrative turnover.

> In several of the sites we studied, teachers who were working with community organizing groups became principals in other schools and were instrumental in developing the next generation of reform educators. Even when they remained as teachers in the school setting, they would often play an important role in keeping up strong school/community connections by "socializing" incoming principals and teachers. In both cases, the assumptions and practice of these teachers and administrators changed as they began to value the community/school connection. In one instance, professionals who considered themselves part of the community organizing effort moved up to central office positions, bringing a community-oriented perspective to the district level. (p. 28)

Alliances with community organizations can not only bring benefits to a school or district, but also bring a form of public accountability lacking in current reforms directed by distant politicians and business leaders. In the absence of real influence on their local schools, poor communities logically turn to the kinds of quick fixes, such as vouchers, that seem to promise a short-term escape from nonresponsive urban schools.

Conclusion

It is increasingly important that school administrators begin to see themselves as advocates for low-income communities rather than paternalistic leaders with a deficit model of urban children and their parents. In the current deindustrialized society that fails to provide a living wage for millions of Americans, low-income parents see fewer options for their children. Two options that loom large in poor communities are incarceration and the military. Zero-tolerance policies in schools and society are viewed as getting tough without having to do the difficult work of building relations and trust with communities. Meanwhile our prisons warehouse 2 million Americans who are disproportionately poor and non-white.

Moreover, the Junior Reserve Officers Training Corps (JROTC) maintains an imposing and growing presence in low-income high schools. Since a decade ago, the military has experienced a well-publicized, post-conscription "recruiting crisis." This crisis has resulted in an attempt to reach young men and women at a younger age—the first and second year of high school. During the past decade, the JROTC budget has more than doubled from $76 million to $156 million.

> The number of JROTC high schools has risen from 1,464 to 2,267, with a 32% increase in enrollment, bringing the number of adolescents enrolled to 310,358. The most recent defense authorization bill (June 5th 2001) called for the lifting of all caps on JROTC expansion, giving

the Corps a green light for expansion into the secondary school system. (Berlowitz and Long, 2003, p. 169)

The post 9/11 patriotic fervor promoted by the government and press has added to the recruitment efforts, and the recent No Child Left Behind legislation contained a little known provision (Section 9528) that requires schools to grant military recruiters access to student information and to school grounds and activities. Schools are threatened with loss of federal funding if they fail to comply.

During the recent Iraq invasion military liaisons and JROTC commanders in some schools were complaining to high school principals about teachers who they felt were not backing the war, leading to suspensions in many cases. At Rio Rancho High School in New Mexico, a student read an antiwar poem that she had written over the closed circuit TV system as part of a regular program promoting the high school's poetry slam team led by English teacher Bill Nevins. The principal received a complaint from the high school Military Liaison who is also a guidance counselor at the high school and the English teacher was suspended, ultimately losing his job. The good news is that the local community, led by active poetry slam supporters and a group of teachers and community members who call themselves the Alliance for Academic Freedom, organized a major poetry event in downtown Albuquerque to raise money for the teacher's legal defense. The event was titled "Poetic Justice: Committing Poetry in Times of War."

As our society becomes more militarized, school administrators' roles will be increasingly defined as enforcers of policies made over the heads of local communities. This principal could have defended this teacher's academic freedom under *Tinker v. Des Moines* which protects the rights of both students and teachers, but chose not to. The pressures on educational administrators to be marketers of schools, image managers, enforcers of testing regimes, and, at the secondary level, military recruiters are profoundly changing their role. Administrator credentialing programs are still teaching a narrow and depoliticized curriculum enforced by new national standards that have little to say about education, much less the pressures on administrators described above. Only through a reconceptualization of the role as one of advocacy and alliance with local community organizing groups can administrators amass some authentic power to counter some of these pressures.

References

Anderson, G. L. (1990). Toward a Critical Constructivist Approach to School Administration: Invisibility, Legitimation, and the Study of Non-Events. *Educational Administration Quarterly*, 26, 1, 38–59.

Anderson, G. L. (2001). Disciplining Leaders. A Critical Discourse Analysis of the ISLLC National Examination and Performance Standards in Educational Administration. *International Journal of Leadership in Education* **4**, 3, 199–216.

Anderson, G. L. (2002). A Critique of the Test for School Leaders. *Educational Leadership*, **59**, 8, 67–70.

Berliner, D. and B. Biddle (1995). *The Manufactured Crisis: Myths, Fraud, and the Attack On America's Public Schools.* Reading, MA: Addison-Wesley.

Berlowitz, M. and N. Long (2003). The Proliferation of JROTC: Educational Reform or Militarization. In K. Saltman and D. Gabbard (eds.), *Education as Enforcement: The Militarization and Corporatization of Schools*, pp. 163–176. New York: RoutledgeFalmer.

Crowson, R. and W. L. Boyd (1999). Coordinated Services for Children: Designing Arks for Storms and Seas Unknown. *American Educational Research Journal*, **101**, 2, 140–179.

Edelman, M. (1978). *Political Language: Words that Succeed and Policies that Fail.* New York: Academic Press.

Edelman, M. (1988). *Constructing the Political Spectacle.* Chicago, IL: University of Chicago Press.

Gold, E., E. Simon, and C. Brown (forthcoming). A New Conception of Parent Engagement: Community Organizing for School Reform. In G. Anderson (ed.), *Policy and Politics*, Vol. 3. In F. English (ed.), *Handbook of Educational Leadership.* Thousand Oaks, CA: Sage.

Harley, B. (1999). The myth of Empowerment: Work Organisation, Hierarchy and Employee Autonomy in Contemporary Australian Workplaces, *Work Employment and Society*, 13, 1, March, pp. 41–66.

Kelly, K. (2002). What about the Incubators? www.emperors-clothes.com

MacArthur, J. R. and B. Bagdikian (1993). *Second Front: Censorship and Propaganda in the Gulf War.* Berkeley, CA: University of California Press.

Moll, L. C., C. Amanti, D. Neff, and N. Gonzalez (1992). Funds of Knowledge for Teaching: Using a Qualitative Approach to Connect Homes and Classrooms. *Theory into Practice*, **31**, 2, 132–141.

Neuman, S. (June 22, 2003). Lynch Now Networks' Objective: The Disputed Facts Don't Matter. The Hype of the Private's Rescue Makes Her Story Rights a Prize. *The Los Angeles Times*, p. 22.

Schattschneider, E. E. (1975). *The Semisovereign People: A Realist's View of Democracy in America.* Hinsdale, IL: The Dryden Press.

Schemo, D. (July 11, 2003). Questions On Data Cloud Luster of Houston Schools. *The New York Times*, p. 1.

School Leaders Licensure Assessment: 1999–2000 Registration Bulletin (1999). Princeton, NJ: Educational Testing Service.

Sergiovanni, T. (2000). *The Lifeworld of Leadership: Creating Culture, Community, and Personal Meaning in Our Schools.* San Francisco: Jossey-Bass.

Shirley, D. (1997). *Community Organizing for Urban School Reform.* Austin, TX: University of Texas.

Shirley, D. (2002). *Valley Interfaith and School Reform.* Austin, TX: University of Texas.

Chapter 7

The Price for "Free" Market Capitalism in Public Schools—or How much is Democracy Worth on the Open Market?

John Weaver

Ever since the Reagan administration published its report *A Nation At Risk* in 1983, the assumption has been that public schools are not performing at an acceptable economic or intellectual level. What is needed, the argument goes, is a good old "healthy" dose of competition that we see in the corporate sector. Parents need choices since public schools are a monopoly and have no incentive to improve the services they render to young people. Since the early 1980s, few people have questioned this conventional dogma. Why do people in the United States immediately assume that the public sector is corrupt while the private, corporate sector is pristine and naturally better? In this chapter, I want to suggest that the first and foremost reason why schools have been declining in quality is the ideology of "free" market capitalism. This ideology has undermined democracy in the United States, reduced the idea of public schooling to job training, limited the intellectual development of young people to economic forecasts, fostered cynicism toward the public sector, and eroded a common sense of community. All of this has been done in the name of potential profit and the growing commodification of humans.

Another premise I want to stress in this chapter is highlighted by the quotation marks I place around the word "free." The reality of the "free" market ideology is very different from any theory about capitalism. For instance, I accept in theory Joseph Schumpater's classic assertion that capitalism is a form of creative destruction. That is, capitalism is constantly rejuvenating itself through the creation of new ways of "doing business" or new ways to create commodities. This creative destruction has enabled corporations to transform themselves from a primarily industrial-based order to one based on global and information technology services. However, make no mistake: the price of transformation has not been free. There is nothing free about capitalism when it pertains to most individuals. The majority of people pay the price of "free" market capitalism through job displacement, economic insecurity, and unfair tax policies. Corporations are the only entities that can proclaim that the market is free. It is corporations that enjoy tax free years, hundreds of millions of dollars in incentives from the federal and state governments, and control of the legal and political systems to make sure they continue to reap the benefits of a "free" economic system.

The unequal distribution of opportunities under the banner of a "free" market economy suggests that when the term "free" is used, it is very limited in scope. "Free" under these terms applies only to the freedom to sell anything (and in some cases anyone). For those entities or individuals who find themselves as the seller, the term "free" implies that they are free from most responsibilities. Under these terms, only the buyer or the person being commodified is responsible. If a person finds himself a victim of unethical business practices (which seems redundant) in a "free" market economy, it is his responsibility to take action. Usually when a person is a victim of an unethical transaction, the traditional response is "buyer beware" or "you should have known it was a scam." These replies assume that the seller has no responsibility to fulfill his or her promise or to meet certain ethical obligations in a rational, free market economy. The "free" market economy is by its nature a predatory culture.

In spite of these socially destructive effects of "free" market capitalism, public schools accept it out of blind faith, intellectual laziness, and financial desperation, without asking at what price the public must pay. A sick dependency has been created between public schools and the "free" market ideology in which public schools continue to adopt the corporate ideology only to create deeper and more serious problems. Public schools, according to Ron Scapp (2003), have accepted the premise that even though the private sector has created many of the problems that public schools face, schools must turn to the corporations for help because "only the . . . corporations can transform education, only they can bring back discipline, accountability, and efficiency, only they can save us (from what they created)" (2003, p. 218).

"Free" Market Capitalism and Predatory Cultures

Peter McLaren coined the phrase "predatory culture" in 1995. A predatory culture, according to McLaren, is "a field of invisibility—of stalkers and victims" that is "fashioned mainly and often violently around the excesses of marketing and consumption" (1995, p. 2). Predatory culture is a new salvation founded on the principles of consumerism. If you possess the means to purchase whatever your heart desires in order to achieve happiness, then you are one of the winners in a predatory culture. However, if you do not possess the means to purchase things or power, then you are a victim. With the power to purchase you can appear on *Extreme Makeover* and remake your life into the fairy tale you have always dreamed of. Without purchasing power, the best you can hope for is to be a famous victim, perhaps making an appearance on *Cops* or any of the other reality television shows that exploit the poor to make millions. No matter where we stand within the predatory culture "food chain," Martin Heidegger's warning, just before he died in 1977, that we are becoming "human resources" is coming true (1977, p. 18).

A predatory culture views everything and everyone as a potential commodity to be sold. As cruel as this may sound, it is important to realize that a predatory culture is not based on coercion but on a willingness to be exploited and used. It is a willful ignorance of dignity and values beyond one's market value. Contestants on *Survivor* are not coerced into participating on the show. Donald Trump did not go up to some unwilling souls and proclaim "You're fired." No one at the networks who produce these predatory shows force people to watch them, although they do spend hundreds of millions of dollars to entice viewers. In a predatory culture, people volunteer to be exploited because in the end their exploitation may be exchanged for riches. *Survivor* survivors get a million dollars, and even the "losers" get to appear on David Letterman's show and all of the morning programs, while "contestants" get other opportunities such as movie contracts, television and commercial appearances, and music contracts. The contestants know they are human resources but they become household, commodified name brands that can earn them thousands, maybe even millions.

Predatory culture is also about a willful ignorance of more important issues that challenge the stability and vitality of democracy. Kobe Bryant was a prime participant in a predatory culture. The only thing that sports commentators and legal pundits on Fox News and ESPN could discuss was who would replace Bryant as the NBA's marquee player. Would it be Melo or Bron? There has not been any serious discussion about allegations of rape or the seriousness of falsely accusing someone. The only discussion was who would step up and be the next NBA star to peddle Nike's latest, sweat-shop made, over-priced sneakers?

Predatory culture is alive and too well. Unfortunately, it is thriving in schools as well. Corporations have flooded the schools with advertising dollars. Contemporary schools appear in a predatory culture like a scene straight out of *Minority Report* where billboards use sensors and lasers to read your eye images and speak directly to you. Students enter schools and see advertisements in the hallways, sports facilities, cafeterias, curricula, and on "educational" television programs. Wherever their eyes focus, there is an advertisement or corporation speaking directly at them: telling them that they are the future of capitalism.

Predatory culture is just as overt in schools as it is in the world of entertainment and official commerce. One does not have to look far to see how corporations are stalking young people. Eric Schlosser cites *Kids Power Marketing* as proclaiming to those corporate predators interested in the young buyers that they can "discover [their] own river of revenue at the schoolhouse gates . . . Whether it's first graders learning to read or teenagers shopping for their first car, we can guarantee an introduction of your product and your company to these students" (2002, p. 52). If this does not provide you with an opportunity to pause and think what the future of teaching looks like, then listen to W. Rossiter, who publishes the *Kids Marketing Report*, when he reveals the desires of corporations within schools: "A successful scheme must have educational value which will help teachers do their jobs, save the schools money, get the children excited, and make parents happy *while still achieving the brand's strategic objectives*" (Kenway and Bullen, 2001, p. 105; emphasis added). Students, teachers, and parents are prey and the predators are more than willing to say and do anything to achieve their brand objectives.

Why do corporations want to solicit so early in a person's life? Research has shown that a person as young as 6 months old can identify a logo, and by the age of two brand loyalties can be created that last a lifetime. Moreover, young people today have a disposable income of anywhere between 6 and 11 billion dollars each year and influence the spending of another 130–160 billion dollars each year. Marketers more than anyone are keenly aware of this spending power and they want as much of that money as they can get. Schools simply have become the site where this competition for this income is played out. Corporations will do anything to get this income even if it means undermining the basic principles of education. The discouraging part of corporations preying on schools is that school leaders, teachers, parents, and young people have not resisted these efforts and assume that anything corporations do for schools is a goodwill gesture.

"Do You Want Fries with That?" Corporate Curriculum Efforts and the End of Thinking in Public Schools

Although it is often not a reality in public schools, there is a long tradition beginning with John Dewey, George Counts, William Harris, and other progressives that one of the main purposes of schooling should be the development of every individual's intellectual potential. Dewey believed that this was best achieved by building on the interests of the child and then connecting those interests to pressing social issues and needs of the day. If schools would develop the intellectual faculties of the child, then young people would be prepared to become active citizens who asked critical questions in a democracy. Other intellectual traditions also believe in the development of the critical faculties of all citizens. The Frankfurt tradition, founded in Germany in the 1920s, is another example. Founded by sociologists like Max Horkheimer, philosophers like Theodor Adorno, and cultural critics such as Walter Benjamin, the Frankfurt School was convinced that if a democracy were to thrive in the world, it had to cultivate critical thinking citizens who demanded justice, created equality, and elevated the intellectual potential of all people. Out of the efforts of the progressives and the Frankfurt School emerged a contemporary movement often referred to as critical pedagogy. Starting with the works of Brazilian educator and activist Paulo Freire and continuing with the efforts of Henry Giroux, Peter McLaren, Joe Kincheloe, and Kathleen Weiler, critical pedagogues believe that a democracy cannot thrive unless public schools encourage students to become actively involved in issues of justice and equality. The connections between critical thinking, democracy, and public schools are threatened by corporations trying to take over the school curriculum and reduce the opportunities of students to create critical thinking powers that permit young people to question the inequities of the world and the many ways in which corporations undermine democracy.

On March 24, 1989, the Exxon oil tanker, *Valdez*, hit the Bligh Reef spilling its contents into Prince William Sound destroying the natural habitat and the economic livelihood of the residents of the Alaskan coastal region. The residents of Prince William Sound successfully litigated against Exxon but have yet to see any of the compensation due to them. Instead of honoring its legal obligations, Exxon continues to appeal the decision hoping to find a sympathetic judge or jury. In the meantime, Exxon has found it within its corporate heart to create a new public curriculum to help students understand environmental issues. Eric Schlosser reports that within this curriculum, students can learn that "fossil fuels created few environmental problems and that alternative sources of energy were too expensive" (2002, p. 55). Of course, one would be hard pressed to convince the residents of Prince William Sound that fossil fuels were not an environmental problem. The goal of this new curriculum is to reconstruct the image of Exxon from a company that is environmentally reckless and uncooperative to one that is environmentally friendly. If students are fed such outrageous claims as fossil fuels are harmless, alternative energy sources are expensive, and teachers do not demand that their students develop critical thinking faculties, then Exxon will never have to worry about paying the damages done to the Alaskan people because it will be easy to find a dim-witted judge or ignorant jury pool who can be convinced that oil slicks really do help the natural habitat, and in the long run, oil spills are beneficial to the economy.

Overt efforts to strip the public school curriculum of any serious intellectual content is not limited to Exxon, now ExxonMobil. Donna Haraway in her book *Primate Visions* (1989) notes that after "years of intense media focus on the system of international oil profits and politics and on the 'energy crisis,' " Gulf Oil company decided to hit the Public Relations tour and associate itself to environmental issues. In its new found support of environmental causes, Gulf concluded that "no thinking person can share in the destruction of anything whose value he understands . . . [Gulf's a]ssociation with the National Geographic Society . . . is only one aspect of [its] lively concern for the environment" (1989, p. 135). Not to be outdone, the American Coal Foundation produced a curriculum that suggested to students that "the earth could benefit rather than be harmed by increased carbon dioxide" (Schlosser, 2002, p. 55). While young people are learning about the benevolence and benefits of oil and coal companies, and learning to evolve because there is less oxygen and more carbon dioxide in the air, teachers need to begin to ask how public schools can overcome the growing crisis of thinking. How can teachers overcome the overt efforts to control how young people think. Corporations are all for critical thinking skills when it comes to calculating how much change a customer gets if they give you $5.32 when their bill is $4.82, or when a customer asks a perplexing question and the employee is able to answer correctly without summoning a manager. However, when it comes to such critical issues as labor rights, livable wages, or general questions of equality in a democracy, corporations much rather feed young people propaganda about the benefits of fossil fuels and carbon dioxide. It is imperative that school leaders, parents, teachers, and students begin to take back the curriculum and revitalize the dreams of Dewey, Adorno, and Freire in order to ensure that a democracy is left in good, active hands as each new generation graduates. Without critically thinking young people, this world will not be for the people, by the people, and of the people but will remain for the corporations, by the corporations, and of the corporations, and individuals will be reduced to "employees" whose most pressing issue in life will be to ask whether a customer "wants fries with that" or not.

"Taco Bell again for Lunch? Cool!" Fast Food Goes to School

I suppose I should start this section with a disclaimer. I grew up outside of Philadelphia which means I love greasy, fatty cheesesteaks and New York–style pizza. Living in Georgia now, I have traveled two and half hours to eat a real Philly cheesesteak in Jacksonville, Florida. I am also a Mountain Dew drinker and have never been on a diet nor will ever become a vegetarian. In spite of these culinary habits, I believe school cafeteria menus are becoming dangerous health risks for young people.

The American School Food Service Association reports "that about 30 percent of the public high schools in the United States offer branded fast food" (Schlosser, 2002, p. 56). Leading the way in this onslaught of the taste buds of the young are Taco Bell, Pizza Hut, and Burger King. Eating at these restaurants once in a while represents no health problems; however, if students eat this food three or five times a week, there are obvious health concerns. Recently, the major news outlets ran a story about a man who ate nothing but McDonald's food for a month. The doctors monitoring his health during that month noted that he gained 25 lb, his cholesterol increased 35 points, and his general health was in complete decline. What are the health risks for young people if fast food is available to them five days a week, one hundred and eighty days a year? Even if you are a conscientious parent, forbid them to eat the fast food, and pack them a lunch, what safeguards are there that young people will not take their own money, dump the packed lunch, and eat Taco Bell everyday? If you are a young person and fast food restaurants target young people with hip, new wave advertising making their food look like the key to being accepted by other young people, what food would you select everyday or most days in your school cafeteria?

The health risks do not stop at fast food. Coke and Pepsi are in almost every school either in machines or at the lunch line. A recent report from the U.S. Department of Agriculture suggested that "consumption of carbonated drinks rose by more than 450 percent, from

10.8 gallons ... on average in 1946 to 49.2 gallons ... in 2000" (Reuters News Service, 2004, p. 1). In the same Reuters news report, doctors suggested that there is a connection between this rise of carbonated drink intake and the rise of esophageal cancer.

How are fast food restaurants and soda companies able to gain such a dominant presence in public schools? Other cafeteria food served in public schools are required to meet the federal dietary standards, but fast food restaurants skirt these standards because they are served à la carte. As a result, fast food is not held to the same standards as other cafeteria food. If a school administrator is strapped for funds and can save money by serving fast food, what will the administrator do? Obviously, they would cut corners and skirt the dietary standards of the U.S. government. Fast food restaurants are interested in serving to young people as soon as possible because they are interested in creating loyal lifelong customers. Once again we see how the desires and profits of corporations are placed above the well-being of individuals. Not only does this issue present a serious health concern for young people if they are eating this food more than once a week, but it also serves as a primary case of how corporations are eroding democracy. By finding loop holes around federal dietary standards, fast food restaurants undermine the ability of the U.S. government to enforce the laws, thereby, eroding trust in the ability of a democracy to function in an effective and equitable manner.

Do as I Say Not as I Do: Corporate Irresponsibility and the Lack of Character Education

Character education is very popular in elementary schools throughout the United States. It teaches children about the need to be honest, patriotic, and responsible. Many corporations are involved in developing character education curriculum, but in reality it is not America's children who need character education. It is American corporations. American corporations are the most irresponsible and deceptive organizations in our society. Harsh words indeed, but let me provide some examples to argue the case.

There is a general lack of responsibility of corporations to the common good. While more and more is being demanded of the individual in the United States, corporations are walking away from any obligation to the public. According to Joe Kincheloe, corporations in 1950 paid 26 percent of the income tax in this nation; by the early twenty-first century, this number is down to 8 percent (2002, p. 110). David Cay Johnston in his book *Perfectly Legal* (2003) suggests that this percentage is even as low as 7.4 percent. To make matters even clearer, from 1996 to 2000, 60 percent of American corporations did not pay taxes at all. How do corporations skirt responsibility while democracy pays the price? One way was revealed on National Public Radio's *Market Place* in March 2004. Corporations conduct phony purchasing scams that on paper look like they are not making profits when they are. One American company recently leased the trolley cars from Dortmund, Germany for 150 million dollars while another leased the sewer pipes from Bochum, Germany for over 100 million dollars. How does one lease sewer pipes? On paper it is easy. The company transferred 150 million to Dortmund officials one day, and the same officials wired the money back the next day earning a hefty 10 million dollar transaction fee for the city of Dortmund. On paper it appears that the American corporation spent 150 million on trolley cars, and therefore, it is not considered part of the profits of the company and is not taxable income. In reality, the company only spent 10 million dollars, making essentially a 140 million, nontaxable profit.

There are other ways by which corporations avoid monetary responsibility for the well-being of communities. In the Cleveland area, city schools lose millions of dollars each year because of tax abatements given out to corporations. Corporations go to the leaders of Cleveland and threaten to move to suburban areas like Berea, Solon, or Cleveland Heights if the city does not grant them tax abatements. In Georgia, the Boeing corporation was thinking of building a new plant near Savannah. Within a few days, the governor's office was willing to provide an incentive plan of approximately 480 million dollars to Boeing. Within weeks of the announcement of this plan, the governor announced that major changes had to be made in the HOPE scholarship program that helps thousands of Georgians to go a state university for free. When it comes to finding money for a corporation, the money is available, but when it comes to the education of young people, the money disappears.

Corporations are excellent at constructing images detailing their benevolence and commitment to people but in reality corporations are irresponsible in their actions. Take R. J. Reynolds as a prime example. In the recent past, R. J. Reynolds ran a commercial in which an employee from one of its subsidiaries, Kraft Food, is highlighted. This good Samaritan works in Milwaukee but commutes to Chicago once a week to help an urban high school student who tries to learn in a school that is underfunded. The student proclaims that he wants to be a teacher so he can do the same thing that his tutor is doing for him. No one would disparage the efforts of this Kraft employee, but one must ask why R. J. Reynolds is putting out such commercials at this time. While R. J. Reynolds is trying to cultivate the image that they really care about social issues and are good "corporate citizens," they are undermining the very things they proclaim to uphold. R. J. Reynolds has spent millions of dollars in research and advertising campaigns to deny that there is a direct causal link between cigarette smoking and cancer. They have not done this because they are good Samaritans but rather because they have spent the last three decades trying to avoid responsibility for their reckless actions in creating an addictive product. Through the Tobacco Institute, R. J. Reynolds and other cigarette makers have conducted their own research not to dismiss the connection between smoking and cancer but to create "scientific evidence" to cast a reasonable doubt that there is a clear connection. Recently, this claim has not been accepted in courts, but to this date corporations are yet to admit that there is a causal link between smoking and cancer. Is this an act of responsibility? Hardly. It is the act of a corporation trying to survive no matter how many people it puts at risk.

R. J. Reynolds is hardly the only company that prefers image making over responsibility. In 2000, it was discovered that Firestone tires on Ford Explorers could explode under certain conditions causing serious harm and death. Because of the potentially faulty tires, there were reported deaths in Venezuela and numerous states in America. In Venezuela, some Firestone executives faced jail time; in the United States, Firestone simply blamed Ford. The shirking of responsibility did not stop there. Instead of taking any responsibility, Firestone did what most American corporations do in a scandal—create a new image. Firestone Tires, a Japanese-owned company, is now called Bridgestone-Firestone.

What do these scandals and cases of irresponsibility have to do with schools? Each time a corporation is able to skirt responsibility for paying local and state taxes, it limits the amount of dollars that governments can spend on public schools and higher education. These cases of corporate irresponsibility also have to do with all citizens, not just teachers and students in public schools. While schools are underfunded, programs for the poor are cut, and troops risk their lives in Afghanistan and Iraq, who is being held responsible? It is easy to make a commercial suggesting a corporation supports public education or supports "our" troops, but when it comes to making a real commitment to individuals, corporations are no where to be found. In a democracy everyone has to be responsible for the common good. If one entity is able to avoid responsibility, then we no longer live in a democracy but in a plutocracy where the interests of a few dominate that of the many.

The Amorality of Corporate America

American corporations are neither moral nor immoral. They are amoral. If lying or creating a new image will earn a corporation more profits, the corporation will do it. If telling the truth will earn the trust of people, corporations will tell the truth as long as it remains profitable. Whatever it takes to earn more profits, corporations will do it. Morality is not the issue. Another reason why corporations are amoral is because they are things. Often the term corporate "citizen" is used to describe the corporate role in society. However, a corporation is a thing: it has no individual rights, it is not an organic being. Given the amorality of corporations, it is important to ask if, we as parents, teachers, and school leaders, want to give corporations any opportunity to influence our children. It is important that young people have a moral foundation to navigate in a world that can be disheartening. It is equally important to recognize that corporations are unable to provide this foundation. The only foundation that corporations can provide is one based on consumption.

There are plenty of examples from the corporate world to make this case. When I have free time from reading and teaching, I usually spend time on the computer playing games, keeping up to date with the latest news about West Virginia University football, or communicating with my friends. Every time I enter one of the various Internet sites, a pop-up ad appears. Most of these advertisements are harmless: "Pick which one is Catherine Zeta–Jones and be a winner." I usually pick the wrong one on purpose, and still win. Imagine that. But there are a few pop-up ads that demonstrate the amorality of Internet corporations. There is one pop-up ad that asks whether George Bush should be elected or not and there is this most offensive one that asks whether or not we should pull our troops out of Iraq. It does not matter which one you select, you always "win." That is, you win the opportunity to spend your money on something you do not necessarily need. The amoral dimension is that corporations take these serious issues and make money off of them. Kenneth Saltman (2003) demonstrates that these pop-up advertisements are not the only examples of amoral corporations. Read the words from this Alta Vista advertisement:

> Who needs elves when you have AltaVista Shopping.com? At AltaVista Shopping.com you can research products you know nothing about: stereos, . . . Pokemon toys, for example. There are 126 different Pokeman characters and over 2,000 licensed Pokeman toys on the market. Only one of them is going to win you most-favored parent status for the coming year. We can help you find out which. (Saltman, 2003, pp. 12–13)

AltaVista has found a way to reduce parenting to a competition to win the hearts of their children. They have turned love into a competition that is no longer unconditional but rather predicated on how well the parents shop for the children. Are these the values we want to teach our young people? Teach young people that Pokemon is the pathway to familial love, and saving 50 dollars is more important than the lives of Iraqis and marines and soldiers? As parents, teachers, school leaders, and students, we need to reevaluate the presence of corporations in our schools. The foundations of our democracy and the values of our young people depend on it.

Re-Creating Public Schools beyond the Corporate Image

If we are to take the encroachments of corporations into public schools seriously, there are at least four things we can do to limit their influence. The first is to recognize that corporations are not evil or corrupt by nature. Rather, corporations have become accustomed to citizens who demand nothing from them. As citizens in a democracy and participants in public schools, we need to rearticulate certain demands and expectations that we expect from corporations. If corporations do not meet these demands, then it has to be known that they will not be permitted to function in our schools or even in our society. Corporations only have the rights that we the citizens grant them. Originally, corporations were granted charters by the colonial powers and later the states. We need to return to this way of "doing business" and revoke the charters of those corporations that do not comply.

Second, there is a need for every school to develop a curriculum that incorporates critical media literacy. Corporations understand the power and persuasiveness of popular culture, and they use it to their advantage to suggest that whatever they peddle is cool and a necessary component in life. Few public schools, however, see the importance in developing a curriculum that permits students the opportunity to critique and understand the impact of the media on their lives. Critical media literacy not only has to give students an opportunity to critique what they see and hear in the media, but also to learn how to use the technology of media to develop their own images and identities. Schools need to incorporate projects into their curriculum that range from the use of video cameras, television cameras, radios, and advertising techniques.

Third, teachers, students, and parents need to reclaim the public sector. Public schools, especially, should not be an area where corporations can take advantage of their financial desperation in order to use public space as a proving ground to create future consumers. The public sector needs to reclaim its legacy as a space where people come to express ideas and participate in a democracy. Corporations should not be invited into this space because they are not citizens of a community; only the owners, managers, and employees of a corporation should be permitted in this space as equal members of a democracy. The public sector should also be reclaimed as a space where individuals can turn to in times of need and come to depend on basic social services that guarantee that our society will remain a humane society.

Fourth, in her brilliant book *Failure to Hold*, Julie Webber has made the case that schools are no longer a place where students can create experiences that prepare them for a life in a democracy. Instead, schools have become targets for containment and consumption. Young people are labeled as potential killers and as a result schools have become more like prisons than democratic incubators. Students are probed, tested, searched, frisked, locked down, and interrogated. When they are treated as humans with individual rights, it is usually as consumers. Corporations are permitted to enter into the schools, train the students on how to become good consumers, and prepare them for a life of satisfying consumption. Consumption is an inevitable part of American life, but when it becomes the driving force that defines a person's identity, there is something missing in the way young people grow up. Schools need to become places where students learn to live in a democracy. Their freedoms need to be restored and the current policies of containment need to be abandoned before another rash of violent reactions surface. If schools do not reclaim their space from corporations and end policies of containment and enforcement, we may be witnessing the creation of a generation of young people who do not know what it means to live a democracy. As teachers and parents, we need to ask what is more important: corporate profits or democratic ideals? Hopefully we will begin to restore our democratic ideals again and end the rise of oligarchic forces.

References

Haraway, D. (1989). *Primate Visions: Gender, Race, and Nature in the World of Modern Science*. New York: Routledge.

Heidegger, M. (1977). *The Question Concerning Technology and Other Essays*. New York: Harper Torchbooks.

Johnston, D. (2003). *Perfectly Legal: The Covert Campaign to Rig Our Tax System to Benefit the Super Rich and Cheat Everybody Else*. New York: Portfolio.

Kenway, J. and E. Bullen (2001). *Consuming Children: Education–Entertainment–*Advertising. Buckingham, England: Open University Press.

Kincheloe, J. (2002). *The Sign of the Burger: McDonald's and the Culture of Power*. Philadelphia: Temple University Press.

McLaren, P. (1995). *Critical Pedagogy and Predatory Culture: Oppositional Politics in a Postmodern Era*. New York: Routledge.

Reuters News Service (May 18, 2004). Too Much Soda may Raise Cancer Risk. MSNBC.com, pp. 1–2.

Saltman, K. (2003). Introduction. In K. Saltman and D. Gabbard (eds.) *Education as Enforcement: The Militarization and Corporatization of Schools*, pp. 1–23, New York: RoutledgeFalmer.

Scapp, R. (2003). Taking Command: The Pathology of Identity and Agency in Predatory Culture. In K. Saltman and D. Gabbard (eds.), *Education as Enforcement: The Militarization and Corporatization of Schools*, pp. 213–221, New York: RoutledgeFalmer.

Schlosser, E. (2002). *Fast Food Nation: The Dark Side of the All-American Meal*. New York: Perennial.

Webber, Julie (2003). *Failure to Hold: The Politics of School Violence*. Boulder: Rowman and Littlefield.

Chapter 8

Bad News for Kids: Where Schools Get Their News for Kids

Carl Bybee

The test of the morality of a society is what it does for its children.
Dietrich Bonhoeffer (1906–1945), German theologian

As parents, teachers, civic leaders, and citizens, we all want the children we work with, as well as our own children, to grow up to be responsible adults engaged with the future of our nation and government. We want them to be good citizens, as well as good parents and good neighbors. And we do not want them to be poor. We want them to be at least economically comfortable, if not financially successful.

And these are all the reasons why we send them to school, to the best schools we can find and create. We believe that learning—the acquisition of knowledge—is one of the greatest stepping-stones that make all of these hopes possible.

When it comes to raising citizens, we expect that our schools will provide our children with the knowledge of our government's ideals, its workings, its formative documents, and its history, good and sometimes not so good, as we have struggled for freedom, justice, and liberty for all. And we also expect that our schools will help our children grow in their awareness of how all of this knowledge connects up with the events of the day—current events—or as most of us adults call it, "the news." What good is a democracy, where the people rule, if the people have little idea about what is going on in their city, their state, their nation, or the world around them?

But there is a great mixed message at work in these hopes. Where once many parents and educators saw the job of raising responsible, engaged citizens *and* well-trained and innovative employees as compatible efforts, these two goals are increasing viewed as in conflict.

After more than two decades of economic turmoil, declining fortunes for the middle class and the working poor, a fear has been unleashed, a fear backed up by reams of statistics, that the link between getting a good education and getting a good job, or any job at all, is falling apart.

Communities, after being battered throughout the 1990s by the aftermath of NAFTA, GATT, the WTO, FTAA, and other international "free-trade" agreements, have seen high-quality manufacturing shipped out of the United States and low-paying service jobs being offered as the primary replacement. Wal-Mart, temporary employment agencies, and the prison/security industry are emerging as the largest private sector employers in the nation. And then these same communities have had yet another economic shock. After a decade of trying to adjust to the "realities" of the new "global" economy by joining the stampede to seduce high-quality, family-wage technology industries to relocate into their industrial parks, often through massive tax breaks and job-training subsidies, these communities have learned a new word in the "global economy" vocabulary: "outsourcing."

Now not only has the United States been hemorrhaging manufacturing jobs, but service jobs and even professional jobs in the technology sector are also being "outsourced." These are jobs that corporations and a now heavily corporate-influenced national government had told Americans were their ace in the hole in the new worldwide game of cowboy capitalism. Our advantage would be in our unique position to hold onto the jobs at the top of the job food chain—the jobs in technology and research—the jobs of the information age. We were told.

With a growing fear about being able to put food on the table and pay medical bills, much less afford rising tuition rates or pay for the basics of K-12 education, parents and educators have become much more susceptible to the arguments being advanced that we must rethink the mission of schools—seeing them less as concerned with citizenship, and more concerned with training our children to be survivors in a world of cutthroat economic competition.

The long-running *Survivor* television series can be thought of as a bellwether, a cultural indicator, of the widespread character of this fear, turning fear into a strange new form of entertainment that preys upon our social and economic insecurity while encouraging us to accept and even cheer on, or at least accept, the return of a new age of social Darwinism. Programs like *The Apprentice* play to this same fear, except that instead of normalizing this brave new world from the point of view of every man and every woman struggling to be the last "man" standing, they invite the audience to identify with the sadistic power of the corporation. For the millions of Americans whose own jobs teeter on the brink of extinction, it allows them to identify with Donald Trump, the corporate lord, tycoon, and entrepreneur extraordinaire whose signature line has become "You're fired!"

The fact is that the message calling for turning our schools more into corporate job-training centers than institutions concerned with promoting and preserving engaged democratic citizens is being delivered by the same corporations that have been involved in or supported sending quality jobs overseas. And these are the same corporations that have managed to reduce their own tax support of our states and federal government, forcing governments to raise taxes on the middle class and working poor or slash education and social services. Yet this vital link has been and continues, for the most part, to be a neglected message in the mainstream news media—a news media that, not surprisingly, is also increasingly dominated by more and more concentrated corporate ownership.

And so as schools refocus, out of fear and corporate pressure, on vocational education (consider my daughter in third grade doing math assignments based on making change at a fast food restaurant), the idea of democracy as people power in the interest of the people is withering away and is discussed less and less, if at all, in the news media and in schools themselves.

At the same time, even as many parents, educators, and civic leaders cling to the dual responsibility of schools to raise democratic citizens and contributing workers in our society, there is a small but important hypocrisy at work.

Still clinging to our belief in the basic goodness of our nation and the moral power of democratic society, we want our children to be informed about the current events of the day and the impact of these events in our neighborhoods and around the world. After the of September 11 attacks, it *was* a heartfelt cry across the country, when the news media asked, speaking for many Americans, "Why?" "Why us?" "How could these attackers, these countries hate us with such a deadly passion?" Yet these are questions that neither our schools nor our news media had given us any way of answering.

And this is the hypocrisy that we must also confront when we begin to question what our children are learning about current events in our schools. The mainstream media have been failing our country in their coverage of issues and events that impact our daily lives in critical ways. This includes their declining coverage of world events at exactly the same time we are being told we live in a global economy to the lack of coverage of the growing divide in the United States between the rich and the poor, to the degradation of work, to the rising tide of poverty, to declining access to health care and shelter, to the withdrawal of funds from states and federal government, to responsibly foster the humane social and economic development of all citizens, and to the continuing rise of unchecked corporate power. *Business Week* is more likely to run a cover story on "Class Warfare" or "Do Corporations Have too Much Power?"—cover stories than they *have* run recently—than our local or regional newspapers, much less our local, regional, or national television news programs.

Our hypocrisy is that, as adults, we are following the news, "current events," less and less every year and we are not demanding that the news industry cover the stories that shape our daily lives in critical ways. Even a growing number of journalists not only admit to the problems but also accept responsibility for what they see as a failure of the press as a democratic institution.

Consider the "Statement of Concern" signed by hundreds of journalists, academics, and media professionals from around the world as members of the Committee of Concerned Journalists:

> This is a critical moment for journalism in America. While the craft in many respects has never been better—consider the supply of information or the skill of reporters—there is a paradox to our communications age. Revolutionary changes in technology, in our economic structure and in our relationship with the public, are pulling journalism from its traditional moorings. As audiences' fragment and our companies diversify, there is a growing debate within news organizations about our responsibilities as businesses and our responsibilities as journalists. Many journalists feel a sense of lost purpose. There is even doubt about the meaning of news, doubt evident when serious journalistic organizations drift toward opinion, infotainment and sensation out of balance with the news.

The cost of the media and the schools in not raising citizens is not just a decline in the level of "civility" in public meetings, debates, and lines at grocery stores, but that the idea of democracy will become more and more of an empty rhetorical term than an active vision of self-governance that places human rights and social justice at the top of its list of moral priorities. North Korea, as well as the United States, considers itself a "democratic" society.

This chapter is about what happens to our children after the school bus picks them up, or they ride or walk to school, or we drop them off at school on our way to work. It is a chapter about the state of current events education in our schools. It is not a chapter about individual stories they might see or read, but more of an essay about where the "news for kids" comes from and the chances they have of making sense of the world they actually live in from this news.

In the first section of this chapter, I take up what we might call the first generation of news products for children, primarily *The Weekly Reader* and *Scholastic Magazine*. The second section addresses the newer players in the news for kids game, primarily television news programs, including the now famous or infamous, depending on your point of view: *Channel One*, owned and operated by the media giant Primedia.

The question I attempt to answer throughout this chapter is "Is this what we need?" Is this introduction to the news of the day in schools enough? Enough to counterbalance their 24/7 exposure to media images, information, and ideas from programs calling themselves everything from "news," to "entertainment," to "reality shows," to "docu-dramas" and "sit-coms," not to mention computer games, based on "real" world events. Especially when the same corporations are involved in producing and integrating their news programs, they are also involved in producing the 24/7 engulfing of children with entertainment media and marketing.

In the summer of 2004, The Television Critics Association voted *The Daily Show with Jon Stewart* the fake news show aired on Comedy Central, for the Outstanding Achievement in News and Information for providing something that the critics felt mainstream news was missing—"a core of truth."

Given the sources of news for children and youth in school and the overt and covert mission of those news sources, what are the chances they are delivering a similar "core of truth" capable of engaging our children in a participatory as opposed to a corporate democratic culture?

The Producers of News for Kids: The First Generation

Up until a little over ten years ago, providing "current events" or "news" for kids was primarily of interest to only two corporate players. For the educational publishing industry, it was clearly seen as a small niche market and was dominated by The Weekly Reader Corporation (WRC), now a subsidiary of WRC Media, Inc. and Scholastic, Inc.

The WRC traces its history back to over a hundred years to its first publication of *Current Events* in 1902 viewed as a newspaper for middle and high school students. Their *Weekly Reader* newspaper was added in 1928, gradually expanding to seven grade-specific editions. First incorporated as the Education Press Company in 1907, it was reincorporated under various names and sold to various owners, including Wesleyan University, in the 1960s to Xerox, in the early 1990s to K-III (now known as Primedia), and sold as part of WRC Media, Inc. to Ripplewood Holdings in 1999.

During the 1991–1999 period when the WRC was owned by K-III, it gained national attention over a study published in the 1995 *American Journal of Public Health*. The study compared the number of antitobacco articles published in *The Weekly Reader* before and after the sale to K-III, which at the time was also the parent company of tobacco giant RJR Nabisco. The results revealed a dramatic decrease (62–24 percent) in anti-tobacco articles and the publication of articles such as "Do Cigarettes Have a Future?" that focused on, without acknowledging it, the tobacco industries' "Freedom of Choice" campaign.

Ripplewood, according to Hoovers Online Business Profiles, is a private equity investment firm established in 1995 with managing "more than $2.5 billion in capital, and invests in automotive retail, food manufacturing, industrial manufacturing, banking, entertainment, and technology. Ripplewood entered the chemical industry when it bought Kraton, the polymers business of Shell Chemicals (part of Shell Oil). In an effort to expand operations in Japan, the company bought Shinsei Bank (formerly Long-Term Credit Bank of Japan) and ailing recording label Nippon Columbia." Ripplewood is making headlines for its success in executing leveraged buyouts in the Japanese market. Ripplewood Holdings, through its ownership of WRC Media, claims to reach over 8,000,000 students, 300,000 teachers, and 60 percent of all public schools in the United States.

Ripplewood appears to epitomize a new vision of what democracy means, a vision born in the 1980s and perhaps most usefully labeled a "neoliberal" model of democracy. Neoliberal democracy is defined primarily by its core belief that "real" democracy is best achieved through the free-market rather than in the messy process of political participation. In this neoliberal vision, the media are no longer considered institutions of public knowledge and debate. Better, in the 1980s FCC Chairman Mark Fowler's view, to consider television as just another appliance. In a statement that shocked citizens' groups fighting to preserve the responsibility of broadcasters using the public airwaves to serve the public interest, Fowler said television was just "a toaster with pictures."

In addition, WRC Media also claims to reach 80 percent of schools with its assessment test division, which has benefited greatly from the neoliberal mantra of accountability for public but not corporate institutions, and 52 million school children through its Lifetime Learning division that specializes in creating sponsored educational materials distributed "free" to teachers. According to Lifetime Learning, they know "how to link a sponsor's message to curriculum standards and create a powerful presence for your message in America's classrooms with informative and engaging materials." WRC Media holdings also include FACTS.com, a Facts on File news service, and Gareth Stevens Publishing, a leading publisher of nonfiction books for K-6 libraries. *The Weekly Reader* division has recently entered into a joint venture with *The Washington Post* to publish *Teen Newsweek*, the teen version of the grownup *Newsweek* with a modest circulation of about 180,000. It is also trying to make the move into television with its partnership with Kids News in the production of the weekly *Eyewitness News for Kids* syndicated program launched in the fall of 2004.

The other long-time player in the classroom has been Scholastic, Inc. The Scholastic Publishing Company was founded in 1920 with the primary purpose of covering high school sports in western Pennsylvania. *The Scholastic* was launched in 1922 to focus on literature and social commentary and targeted at high school English and history classes. Over the next 80 years, Scholastic added specialty magazine after specialty magazine, along with grade-appropriate editions of *The Scholastic*. By 2003, Scholastic was publishing over 35 classroom magazines with a readership, claimed by corporate reports, of almost 23 million students in K-12. During the years the corporation added a national writing awards contest, an arts awards contest, and moved into the book publishing industry, where it now ranks among the top ten publishers and distributors of children's books in the world, including the wildly profitable Harry Potter series marketed through nearly every division of the Scholastic corporation.

In the 1940s, Scholastic moved into the school book club business, at first providing a unique distribution outlet for books published by other corporations and later serving as enormously powerful distribution system for its own titles. The various school book clubs, now numbering 11, were coupled with fund-raising efforts by schools, which enhanced the value of the Scholastic brand in the schools, and netted the Scholastic corporation an enormous army of parents and teachers, working as volunteers, to aggressively market Scholastic titles.

In the 1950s, Scholastic began to go international, first Scholastic Canada and then following with the United Kingdom, New Zealand, Australia, Mexico, India, and Hong Kong.

In the 1960s, Scholastic's "Lucky Book Club" offered Norman Bridwell's *Clifford the Big Red Dog*, a book that has launched a brand empire for Scholastic, which now includes books, television series, and licensed merchandise. In 1997 there were more than 68 *Clifford* titles with 70 million copies in print. The corporation also began moving into the core curriculum market in the 1960s, which followed with the development of television, feature film, video, computer software, and online services.

In the 1970s, Scholastic Productions was created to "extend company franchises across multiple media." Scholastic Productions was renamed Scholastic Entertainment in 1998, producing children's programming and multimedia materials and serves as a worldwide licenser and marketer of children's entertainment properties.

In the 1980s, Scholastic moved into the book fair business, eventually acquiring Great American Book Fairs, and emerging today as, according to the corporation, "the largest children's book fair operation in the United States," creating itself a formidable gatekeeper to entry into children's book publishing. The corporation also began to move more aggressively into producing teacher resources, promoting new technology in the classroom, and establishing its own New Media division to produce educational software. *The Magic School Bus* book series and brand was launched, along with the *Baby-Sitters Club* book series and brand. By 1995, Scholastic had partnered with Columbia's Tri-Star Home Video to produce a *Baby-Sitters Club* movie.

The corporation continued its move in the 1980s into specialized magazines and books for teachers on educational practice and into core curriculum development. In 1992, Scholastic released the first in the series of *Goosebumps* books by R. L. Stine. By 2003, its marketing, promotion, and distribution system had moved the series, according to Scholastic, into "the number one children's book series of all time, with over 167 titles and 215 million books in print." In 1993, Scholastic launched its first web site on AOL and relaunched it as a stand-alone site in 1996.

In 1994, the Scholastic produced *The Magic School Bus* premiered as a weekly series on the Public Broadcasting System. In 1998, Scholastic sold the series to Fox Kids Network, stepping up their expansion out of the classrooms and public broadcasting sector into private sector partnerships.

For Fox it was an opportunity to buy a cheap set of children's reruns that would help them meet the letter, if not the spirit, of the Children's Television Act of 1990, then receiving some regulatory attention from Reed Hundt, President Clinton's appointed FCC chairman. Over the next few years, Scholastic partnered with Microsoft, Warner Home Video, DreamWorks Interactive, Fox Home Video, Paramount Pictures, Oprah's Book Club, Miramax, Nickelodeon, HBO, *The New York Times*, NBC, John F. Kennedy Center for the Performing Arts, New Line Cinema, New Video,

The National Institute on Drug Abuse, The History Channel, UPS, The Advertising Council, The Declaration of Independence Road Trip, DreamWorks, The Learning Channel, PBS, MSNBC.COM, and the National Football League. In the summer of 2003, Scholastic Inc. launched a 1 million dollar marketing venture for the McDonald's Corporation to specifically reach children under the age of 6 through a mailing to the nation's largest 22,000 preschool and kindergarten classrooms. Scholastic's vice president of business development said, "Programs like this help us refine the Scholastic database because we get feedback from the teachers. The feedback tells us that here's a teacher open to receiving custom programs from us." According to Direct Marketing Business Intelligence, "McDonald's intent is to associate its mascot, the clown Ronald McDonald, with preschoolers' emotions toward learning to read, and to raise awareness of Ronald as a brand icon among kids."

Scholastic's unique marketing system works furiously to maintain its "educational" brand identity and to solidify its marketing and promotion stronghold in the schools. This is done through the continued production and distribution of educational materials for students, teachers, and administrators as well as continuing to link these in-school promotions to school fund-raising at a time when declining school operating revenues are increasing schools' needs for external funds. And what more pro-education way to raise funds for schools than to sell books to students and parents and encourage the latter to buy books (Scholastic distributed books) for their school libraries. Scholastic also continues to maintain its "educational" brand identity by sponsoring pro-education, pro-literacy events, such as becoming a continuing sponsor of The National Teacher of the Year Award beginning in 1995.

Launching shows like *The Magic School Bus* on PBS also helped build its "education" brand identity while garnering awards from parent and industry groups alike. The PBS "educational seal of approval" and the low entry costs to airing a program on PBS, then set up opportunities for moving into much more lucrative commercial slots, like on the Fox Kids Network. By 1997, Scholastic was well aware of its powerful position in the in-school marketing world and established its Scholastic School Group, "a sales, marketing and promotions unit that combines IPG, Magazines, and Supplementary Publishing . . . divisions into an unified group that sells to the institutional market."

In the same year, the marketing world was beginning to take note of the Scholastic integrated marketing machine. Scholastic Productions received the 1997 Licensing Industry Merchandizing Association Award for "Licensing Agency of the Year," naming

Goosebumps the "License of the Year." Scholastic Entertainment was "awarded the Reggie Award by the Promotion Marketing Association for the innovative Goosebumps/Pepsi/Frito-Lay Consumer Promotion." The *Goosebumps* license also extended to include other fast food and junk food vendors such as Taco Bell and Hershey's chocolate in their "Reading is a Scream" promotion. The growing childhood obesity epidemic had not quite made it on to front pages across the nation.

Scholastic continued to win awards for quality children's products, such as the 1998 recognition by the NAACP for their *I have a Dream* children's book while acquiring companies like QED, which specializes in marketing to schools, selling, according to their promotional material, "exclusive school databases, school demographic data, and research and database marketing services." Scholastic, in 2001, also moved into the retail store business where Scholastic products can be seen, promoted, and sold. In 2002, Nelson B. Heller and Associates, a "leading business-to-business publisher of educational market newsletters and market research," formed a partnership with Scholastic under their Marketing Partners division.

While clearly not a megamedia corporation on the scale of Disney or Time Warner, Scholastic with annual revenues in the 2-billion-dollars-a-year range shares many similarities with the operating system of the Disney Corporation: synergistically linked divisions, cross-promotions, careful maximization of licensing opportunities, multiple outlets for single products and brands, and careful attention to managing brand identities, from the corporate name to the vast array of corporate products. Of course, the key difference is that where Disney's original base was in children's entertainment, Scholastic's base has been and continues to be in children's "educational" products—particularly spearheaded by its "news for kids" division, although both corporations have dramatically blurred the boundaries between education and entertainment.

However, given Disney's entertainment base, when it signs a cross-promotional marketing deal with McDonald's, for instance, it stands to draw much less critical attention than when, say, Scholastic served as a key sponsor of the Golden Marble Awards, a marketing industry award ceremony celebrating marketers who have found the latest and most innovative methods for selling to children. Or when Scholastic produced *Geoffrey's Reading Railroad*, a program to "encourage children and families to spend more time reading together," for Toys "R" Us.

In the Consumer Unions' 1998 Report, "Captive Kids: A Report on the Commercial Pressures on Kids at School," they gave the program its worst rating, labeling it "highly commercial." While students can earn points for class prizes, the prize for reaching the book goal was a free book to be picked up at Toys "R" Us and the opportunity to enter a sweepstakes to meet Toys "R" Us's branded characters. As the Consumer Union reported, the sponsor's name and trademark were on all the materials and the Toys "R" Us branded character was incorporated into the title. In addition, materials contained coupons only redeemable at Toys "R" Us, prize points could only be redeemed at the store, and the entire program was organized around launching a new Toys "R" Us marketing line, Books "R" Us. Most of this criticism, however, did not make it into the mainstream media and, for the most part, has done little to alter Scholastic's overall "edutainment" marketing strategy.

Before we consider the implications of this first generation of "news for kids" providers, we need to briefly consider a third, although somewhat different player in the classroom: Newspapers in Education (NIE). What makes the NIE program different from the WRC and Scholastic, Inc. is that it is an association and not a single corporation and that it is not engaged in direct marketing in the schools. The use of newspapers in the schools had been encouraged and supported by a number of individual news organizations throughout the 1930s and the 1940s both as a civic contribution to the community and as a training ground for raising new generations of newspaper readers. In the aftermath of World War II, the beginning of the cold war and the Commission on the Freedom of the Press' 1947 report, "A Free and Responsible Press," these concerns, and particularly the idea of the democratic responsibility of the press, were taken more seriously.

In 1957, a "National Newspaper in the Classroom" program was initiated that eventually was renamed the NIE Program and brought under the umbrella of the Newspaper Association of America. The program provided support for newspapers in organizing their own local programs and conferences and seminars showcasing exemplary efforts. The program grew throughout the last decades, establishing a service-based web site, and providing support to over 700 NIE programs around the country. In the last 20 years as newspaper readership has continued its slow decline with adult audiences and its precipitous decline with young adults, the program has become more oriented to a news marketing dimension of its mission, introducing marketing vocabulary such as "news as product" and "brand building." And newspapers have become much more interested in moving out of simply providing newspapers and lesson plans to schools, into more systematic marketing strategies. As one newspaper industry analysis puts it:

> NIE is a good first step, but relying on it exclusively has its risks. Can you imagine Coke, Pepsi or General Mills entrusting product introduction, promotion and support to teachers who have no stake in its success? Would periodic meetings with advisory councils of 10–12 prospective customers be enough for their product development programs?

Reuters and cable giant TCI did not think so when they joined forces to create *Ingenious*, a new 70-person company that is trying to build on TCI's base of providing an educational service to 25,000 schools and homes. The new venture committed 25 people and 6 months to developing *What On Earth?*, a daily multimedia news journal that was eventually beta tested by 300 schools and evaluated by 500 more. Available for 149 dollars per year, plus network fees for each computer site, *What on Earth?* is produced by four, three-person teams of journalists, teachers, and multimedia authors. . . . And for interactivity there's a web page and carriage on the Microsoft Network.

Although Reuters expects *Ingenious* to be only a modest business, its goals are clear: to gain immediate access to the growing educational market plus increased brand recognition that can be transferred to the consumer market. Most important, we believe, is its serious commitment to achieve those goals—which is more than you can say for most of those 700 newspapers that do not even bother to participate in the NIE program.

While the *What on Earth?* series never took off, it did reveal the desperation of the news industry that is left with only "news" as product, a niche product that is continuing to lose its appeal.

The subject of the "future of news" is a hot topic across the spectrum of news media industry outlets, from television news, to newspaper news, to magazines and cable. The newspaper industry is currently the most shaken up because they have lost the most ground in audience size and are technologically the most distant from becoming part of the digital media revolution.

From the University of Southern California's Annenberg's *Online Journalism Review* with its "Future of News" initiative to the media think tanks ranging from the Poynter Institute to the Project for Excellence in Journalism, discussions of what is in store for the news industry are hot. However, most of the discussions taking place are about economic survival, not the place of a free press and an informed citizenry in a democratic society—and certainly not about the very meaning of "democracy" being transformed from "people power" to "marketplace power."

The debates that are taking place about youth and news are more focused on hooking kids on the importance of following the existing news product (be good, bad, or useless) and instilling in them a sense of news brand loyalty. Issues like the relationship between information, critical understanding, political awareness, and self-governance are almost nonexistent. In a recent interview, kid anchor Haley Cohen of the *Weekly Reader*/Kids News *Eyewitness News for Kids*, daughter of celebrity mother CNN anchor Paula Zahn, was asked if she liked watching the news. "Well, *I* find it interesting to watch," she said, "but sometimes you just have to turn it off, because it's very harsh, very blunt and to the point, and it's all, I don't know—I turn on the news sometimes and it's really sad sometimes because there's practically no good news anymore, and that can get depressing, and we're trying to balance the hard news with the fun stuff so it's easier to watch."

Where does this leave us in this quick review of the first generation of producers of news for kids? What becomes clear is that this is not an unfolding story of ever-more kid-friendly and civic-minded initiatives for raising democratic citizens. It is not a story of carefully reviewing the growing body of knowledge about the necessary skills of citizenship and their relationship to developing an appropriate political consciousness among young people. It is not a story about the quest to help young people understand the changing meaning of democracy and its relationship to the press, much less to themselves. It is certainly not about challenging the ideology of consumerism over caring or the connection between social justice and free speech and media accessible to all people and views.

It is the story of small-time educational marketers, the WRC and the Scholastic, Inc., that have grown into large, diversified media corporations. These were media corporations that began primarily as vendors of a product that was seen as relatively unique in the educational realm, "news for kids," but without the historical appreciation of the link between democracy and the press. The question of "news" for kids was as likely, if not more, to be understood in connection with the development of literacy skills than civic skills. It is also a story of two corporations that held near monopolies within the educational realm over the production and distribution of this product and were able to build successfully on their unique product and foothold to expand their corporate presence in the schools and classrooms. In a sense the story can be read as one of the first instances of corporate "cause marketing," using the link to a valued and socially positive action, in this case bringing current events to children, with ever more sophisticated educational marketing integration.

Scholastic, in particular, recognized the value of its positive branding in the educational sector to bring a wide range of additional educational and quasi-educational products through its school distribution pipeline into the classrooms and homes of students. With the rise of the neoliberal vision of democracy in the 1980s and the consequent attack on the public schools as examples of the failures of the public sector in terms of management and results, corporations like Scholastic and WRC Inc. were well positioned to move into a market that had been historically off limits. The "news" arm of these corporations that had been modestly successful in economic terms were now able to continue to serve, along with calls for expanding literacy, as both products and covers for diversifying educational marketing, developing testing, and assessment systems as corporate products, and providing educational marketing services to other corporations looking to break into the classroom market.

The business models driving these corporations, the ideology of glorifying technology over teachers and pedagogy they celebrate and sell, the leadership roles they play in the growing commercialization of the classroom, the blurring of the line between education and entertainment marketing through their licensing efforts, and their ability to exploit not just schools but also publicly funded public broadcasting in subsiding consumer recognition of their corporate brands, all draw on and serve the growing acceptance of the equation of neoliberal economics with neoliberal democracy. In their corporate practices, they gave concrete practice to the promise that deregulated marketplace efficiencies joined with science and technology, privatization and unleashed individual choice would accomplish what the cumbersome institutions and processes of democracy could not.

The news industry, increasingly a smaller and smaller part of the growing media industry, was also choosing or being forced in the interest of economic survival to accept the new neoliberal vision of consumer-based citizenship.

By the middle of the 1990s, for both *Weekly Reader*, now a subsidiary of WRC Inc., which became a subsidiary of Ripplewood Holdings, and for Scholastic, Inc., their web sites have become portals for linking student readers of their no-advertising news magazines, to advertising and marketing rich environments. For instance, in November 2003, *Weekly Readers*' home page for kids contained links ranging from sponsored programs tying students into QVC, the home shopping cable network, Nike, IBM, Kleenex, Charles Schwab, General Motors, Polaroid, KidsSmart Educational Technology, and Six Flags Theme Parks. In addition, students could enter the "Operation Tribute to Freedom" essay contest, a tie-in with the

Defense Department's "Operation Tribute to Freedom Program" encouraging "Americans to show appreciation and support for military personnel fighting the war on terrorism in Iraq and Afghanistan." The grand prize would be a trip to Washington, DC to be "a kid correspondent and report from the Pentagon."

Scholastic, Inc.'s November 2003 web site home page for kids, promoted its full range of products from *Harry Potter* to *Captain Underpants*. At the same time, it included a range of activities from contests to quizzes that implement the goals of their Scholastic InSchool Marketing Division. The division specializes in "the development and distribution of branded in-school and consumer marketing programs. Programs include: Brand Awareness, Direct to Home Marketing, Retail Tie-In, Consumer Loyalty, QSR Programs, One-to-One Marketing, Public Relations Tie-Ins, and More."

The Producers of News for Kids: The Second Generation

Channel One

The early efforts of the WRC, Scholastic, Inc., and the NIE programs were at best, low to invisible, efforts on the part of corporations to bring news to young people. Even while the efforts of WRC Inc. and Scholastic, Inc. became more aggressive and diversified throughout the 1980s, and the NIE stepped up their efforts to stop the hemorrhaging of the youth news audience, overall they had done outstanding jobs of maintaining their brand identities as primarily small-time, educational do-gooders.

In 1989, Chris Whittle blew the top off this corporate news game in the schools with the launch of his ambitious daily in-school video news program *Channel One*. Over a decade later, of waxing and waning controversy, claiming to reach 40 percent of all middle school and high schools across the United States and 8 million teenagers, *Channel One*, currently owned by megamedia giant Primedia, is still in operation. Most parents, most adults, have never heard of the program. At a recent local school board meeting in my university hometown, a community that prides itself on its commitment to education and children, I asked the school board members if *Channel One* was operating in any of the district's schools. Not one board member knew. Several had never heard of *Channel One*. The superintendent of schools said authoritatively it was not operating in the district although he was wrong.

In 1989, Chris Whittle had thoughtfully read the intersections of a number of key forces that made for an outstanding business plan.

Schools had been under neoliberal attack for nearly a decade endlessly described as inefficient, wasteful bureaucratic dinosaurs harboring lazy, do-nothing, unaccountable public employees. At the same time, the combination of tax cuts for corporations and for the wealthy and a historically unprecedented peacetime military buildup had coalesced into the "starve government" neoliberal politics of the 1980s forcing cutbacks in social programs making up the nation's "safety net" and decreasing tax revenues for schools and all levels of government. The Reagan administration preached fiscal austerity and tripled the national deficit. At the same time, the new industrialization of information—computer technology was exploding—with its attendant ideology that whatever the problem was, technology, not equitable public policy, smaller class sizes or social justice, was the answer. Media deregulation was in full swing. The call for privatization of all things public was also being pushed with increasing stridency by the newly founded collection of neoliberal think tanks, from the Heritage Foundation to the Enterprise Institute. Simultaneously, corporations were just beginning to appreciate the dazzlingly successful "cradle to consumer" marketing strategies of the McDonald's and Disney corporations. And their own research divisions were just beginning to crunch the numbers on how quickly the purchasing power of children and their ability to influence family purchases were growing. Youth between the ages of 12 and 19 spent 155 billion dollars of their own money in 2001 and influenced family purchases of nearly 200 billion dollars.

Chris Whittle, who founded Whittle Communications in 1970, had already discovered the growing importance of niche marketing—finding those small, but enticing, advertising-free zones overlooked by the major media, but increasingly lucrative. Lucrative, first, because they were untapped advertising territory, and second, because they had the potential to offer high-quality viewer demographics—for instance, producing specialty magazines at no charge for physicians' offices, in order to sell advertising space, and advertising posed as editorial content to bored patients.

In the case of *Channel One*, Whittle also had the extremely successful print prototypes offered by Scholastic, Inc. and WRC Inc.—getting one's marketing foot in the door with what appeared to be an obviously positive educational product: the news. Why not offer cash-strapped schools a free daily video newscast, produced specifically for the middle and high school market? And to make it a deal that most poor schools literally could not refuse—provide all of the technology, the television sets for every classroom, the video recorders for every classroom, and the satellite receiving stations for free. Principals were merely required to sign contracts committing their schools to air the 12-min program at least 90 percent of the regular school days, to 90 percent of the classes. A school, for allowing 12-min of high-quality, kid friendly news to be piped into their school could receive, on loan, as much as 50,000 dollars worth of perhaps otherwise unaffordable video equipment. The only catch was that each daily newscast would contain 2 min of television commercials that the Whittle Corporation could sell to advertisers desperate to address this precious but notoriously difficult to reach demographic group.

As Ed Winters, one of the co-founders of *Channel One* with Whittle put it in 1997, "Marketers have come to realize that all roads eventually lead to the schools." Whittle would be there waiting for them when they showed up.

The story of *Channel One* and the continuing battle over its presence in classrooms has been well documented in a number of articles and books. The importance of the *Channel One* story to our discussion comes from three related lessons it can teach us about the complexity of the kids, corporate news, and citizenship question. First, the *Channel One* story tends to be framed by critics as the leading example of the commercial invasion of public schools, confusing and breaking down the border between private enterprise and the public sphere. Second, while there some considerations of the quality of *Channel One* news, as opposed to its blatant commercialism, the question of what is appropriate news, both for children and adults tends to be neglected. The third lesson is the implication that the problem of *Channel One* could be solved by its removal or its replacement with a commercial-free alternative youth news program. This neglects the deep integration of structure, content, and marketing in the construction of these programs.

Let us take each of these lessons in turn and consider them in more detail. The public sphere versus privatization argument is clearly critical, yet the very idea of a public sphere is an extremely abstract concept, even for public employees such as schoolteachers. Publicness at its best is most likely to be associated with public spaces like parks, which tend to be used for private consumption and enjoyment, not as meeting grounds for forming associations. More often, public spaces are likely to be perceived as threatening and lacking security. Malls are safe. Public parks are not. And none of this even remotely connects to the idea of "publicness" as a means by which individuals form a sense of themselves as a group with collective interests. All of which is to say that the key distinction between privatization versus public sphere may best be recast and linked to understanding different models of democratic life, since the concept of democracy, while dangerously emptied of meaning, still maintains a positive resonance.

The second lesson of *Channel One*, the focus on its commercialism in the schools rather than on its commercialism in relationship to its news product creates a different kind of problem. Most of the concern about *Channel One* has come from the critical educational community and the growing critical media literacy community and not from the press community, although exceptions exist.

The mainstream press may be willing to do an occasional story on *Channel One* and the resistance directed against it, but these stories will tend to be framed largely from an educational point of reference. To question the intersection of advertising and news would be to question the very foundation of their own industry. Critically oriented news media scholars have tended to neglect the significant political socialization work taking place in this world, nearly invisible to adults. The result is that when we see organizations or institutions address the problems of youth political apathy and cynicism, and the issue of civic re-engagement, there is almost no attention given to the political and ideological crisis of meaning taking place in mainstream journalism, much less "news for kids." These reports call for bringing more "current events" into the classroom with what seems to be no clue as to what exists in the landscape of "current events" available to young people.

This brings us to the third lesson we can learn from *Channel One*. For many educators and media activists concerned about the impact of *Channel One* in the schools, the most common solution advocated is to drop *Channel One*, now owned by media megagiant Primedia, and replace it with the commercial-free option of *CNN Student News* offered by Turner Learning, the Education Division of CNN and Turner Broadcasting, a Time Warner subsidiary. But as indicated above, this simple strategy has its own problems.

Nineteen eighty-nine was also the year in which the Cable in the Classroom initiative was launched by the cable industry to offer free cable access and commercial-free cable programming to schools. *CNN Student News* (then called CNN Newsroom) was developed and offered under this program. In this interesting liberal turn of neoliberal thinking, one can see both a recognition that there might be something valuable to offer children by excluding commercials, but at the same time a failure to consider that if the programming were initially developed for commercial purposes, or to explicitly carry commercial messages, that the logic of this motive might be imprinted on the programming itself. Further, in considering that the Cable in the Classroom initiative was more than just civic generosity, it must also be taken into account that it served as a critical dimension of marketing through the creation of brand awareness and positive brand associations. Further still, it needs to be considered that the Cable in the Classroom initiative grew out of, in part, strategic efforts on the part of the cable industry to ensure that there was no regulatory backlash to the 1984 Cable Act that largely expanded the scope of private enterprise in the field of cable. By the late 1980s, with rising cable rates and failed programming promises, this had been a distinct possibility.

A Brief Survey of the Newcomers to U.S. Kids' News Production: More Bad News and Some Glimmers of Hope

How does one even begin to look for sources of news specifically produced for kids? One can "Google it" or "Yahoo! it," for that matter. Yahoo! actually provides a subcategory "News for Kids" organized under "Society and Culture—Cultures and Groups—Children" with 13 highlighted sites out of 66 overall listings. The 13 highlighted sites included children's news sites from Canada, Britain, and Australia. *Weekly Reader* and its magazine produced in partnership with The Washington Post Company, *Teen Newsweek*, did not make the list. *The Washington Post's* web site for kids, KidPost, did. Scholastic, Inc. and the *New York Times* made the list, Time for Kids and *Channel One* did not. Yahoo!'s own "Yahooligans! News" claiming to feature "top stories, popular news, photos, and weird news," not surprisingly made the list. ABC News Network's ABC *News 4 Kids* made the list, even though the site had not been updated in two years.

Overall, as an entry point to "news for kids," the Yahoo! site or even a Google search is a librarian's nightmares. No evidence of selection criteria, no indications of age-level appropriateness, and in many sites and linked sites, such as Kidsnewsroom.org and PencilNews—News for Kids, there was no evidence of who the operators or sponsors of the organizations were. For instance, on the Kidsnewsroom.org site, the only clue as to who was behind the site was a mention in their "About Us" section that "One of our premier sponsors, Top Centre Tickets, deals in premium sporting tickets, such as Super Bowl Tickets." Click on "Super Bowl Tickets" and one is immediately connected to the Top Centre Ticket site. Although no identifying information is provided, *PencilNews* is a production of the MSNBC cable channel that is a subsidiary of General Electric.

If one seeks to create an inventory of "News for Kids" sources from organizations or institutions that make visible their funding sources, their boards of directors, their organizational structures, and their operating systems that are "transparent," in the neoliberal jargon of the day, the options are few. The only press-related organization providing recommendations is Cyberjournalist.net that is a project of the Media Center at the American Press Institute. Its mission is to help build "a better informed society in a connected world." The Media Center is funded by a range of foundations, major media corporations (including *The New York Times, the Wall Street Journal*, CNN, Time Warner), corporations (including the Rand Corp., IBM, Microsoft, and Ford), a number of media institutes, and several major universities. Their list roughly mirrors the Yahoo! list, including Yahooligans! News, Scholastic News Zone, Weekly Reader, Time for Kids, KidsPost, and Britain's CBBC Newsround.

The American Library Association, in their "Great Web Sites for Kids," features a "News and Current Events Section." Eleven sites are featured. Four are for standard adult news web sites: ABC, CNN, MSNBC, and PBS's the Online News Hour. The only news site that was specifically listed as designed for kids was *Time for Kids*. Apparently they were unaware that CNN offered *CNN Student News*, and the *Online News Hour* has created a special section called "Extra" for students. The National Council for the Social Studies provides no evident guidance to teachers seeking sources of age-appropriate news for their students.

Apparently the only child advocacy organization making recommendations for "news for kids" is Children Now. Children Now, founded by a Stanford law professor in 1993, recognized early on that to meet the challenges being faced by children in terms of health, education, and creating healthy families would also require a focus on the media dimensions of raising children including the images consumed both by children and policy makers about the status of children in the United States. The Children Now list of recommended kids' news sources contains again the familiar *New York Times* Learning Connection, Yahooligans! News, Scholastic News Zone and Scholastic newsmagazines, Pencil News, Time for Kids, and Canada's *CBC4Kids*. Interestingly enough, it also mentions Nickelodeon's *Nick News* and *Sports Illustrated for Kids* as top choices. The Children Now list comes as a resource guide at the end of an extensive downloadable guidebook titled "Talking With Kids About the News," part of their "Talking With Kids About Tough Issues" series.

Once we complete our inventory of "new sources for kids" list recommended by organizations or generated through popular web search engines, we first notice that it's a short list, then we notice that they are almost all products of the media megagiants, and finally we notice that even on this short list, there is a remarkable degree of joint corporate ventures.

From the recommended lists mentioned, Time Warner accounts for *CNN Student News*, Time for Kids, and Sports Illustrated for Kids. Time Warner has also partnered with Weekly Reader to produce a weekly children's news program *Eyewitness Kids News* that carries advertisements and is syndicated to over 150 broadcast and cable channels around the country. Viacom owns and produces *Nick News*. Primedia owns and produces *Channel One* and, until 1999, owned the WRC Inc. When Primedia owned WRC Inc., it created a joint venture with the Washington Post Company and its subsidiary *Newsweek* to produce *Teen Newsweek*, still in publication, but now, as mentioned above, owned by Ripplewood Holdings. The New York Times Company produces the New York Times Learning Connection web site and has a joint venture with Scholastic, Inc. producing the New York Times *Upfront*, a newsmagazine for teens as well as a companion web site. In 2003, Scholastic.com has also signed a joint agreement with General Electric's MSNC.com. This agreement features Scholastic.com on the MSNBC.com web site and created a "co-branded" web site scholastic.msnbc.com. "The agreement between Scholastic.com and MSNBC.com creates an exciting, robust and comprehensive news resource speaking to a broad audience, from kids in the classroom to adults in the global consumer space," said Donna Iucolano, president of Scholastic. As part of the agreement, Scholastic.com will receive a vast set of interactive features "on numerous topics, including science, technology, space exploration, and social studies. These will appear on Scholastic.com's 'Online Activities' and be used by teachers and students in schools." Parent corporation General Electric, heavily involved in science, technology, space technology, and military research and production should be pleased with this arrangement. The Disney corporation has for the most part stayed out of the news for kids business with the exception of the now defunct ABC News 4 Kids web site venture.

We can also round out our list of top mainstream sites by looking at what the PBS has to offer. It took the U.S. government nearly 40 years longer than any Western European country to launch a public broadcasting system. Almost as soon as the Public Broadcasting Act was passed in 1967, it was entangled in political and economic wars, ranging from unstable funding, underfunding, to the efforts of the Nixon administration in the early 1970s to slash funding and increase White House control over the system, and the successful efforts by the Reagan administration to severely cut federal funding for public broadcasting in the early 1980s. All of which is to say that one's meager expectations regarding the support of current affairs programming for youth are well met.

The Public Broadcasting Corporation, under constant defunding threats from the federal government, under constant attacks from conservative groups for their supposed left-wing biases, and under the growing need to win corporate sponsorship has continued to move to the right and has been exceptionally timid, with a few exceptions, in its efforts to produce hard-hitting, critical, and contextual news coverage for adults, much less for youth. PBS currently does not produce any "current events" programs specifically for young people, unless one would want to count PBS's efforts at "reality" shows such as *Pioneer House*. However, PBS has supported web site development of teacher support materials, and student resources for three of their news or documentary programs: *The News Hour with Jim Lehrer*, and *Frontline*. The web sites associated with each of these three series provide in-depth teaching materials and resources for teachers and students unlikely to be found anywhere else in the world of corporate produced kids' news. The audiences for these programs and materials are of course, tiny in contrast to say *Channel One*, *CNN Student News*, or Scholastic, Inc. news products.

To summarize and to fill in a few blanks, the main terrain of news for kids contains a short list of video news, news magazines, and web sites typically tied to either the same video news programs or the same magazines.

In the area of video news for youth, there are currently six regular programs, *Channel One*, *CNN Student News*, *CNN Presents Classroom Edition*, *Nick News*, *Nick News Special Edition*, and *Eyewitness Kids News*. *Channel One* is the only program being directly distributed into the schools.

Time Warner produces *CNN Student News* that is cablecast 5 days a week and *CNN Presents Classroom Edition* that cablecast as an occasional special feature series. Both, while intended for classroom use, are cablecast in the early hours of the morning for teachers to prerecord and bring to their classes. Each 10-min daily episode of *CNN Student News* is also available through video streaming on the *CNN Student News* web site. At the web site, teaching resources to be used in conjunction with the programming are provided, along with a range of Time Warner tie-ins and advertisements. Turner Learning claims that its Student News program is being used in 18,000 classrooms on a daily basis across the nation. If this is true, the audience base is somewhere between 300,000 and 600,000 students each day—considerably fewer than the 8 million claimed by *Channel One*. An attempt in the spring of 2002 to include advertising in the programming, in order to turn what Time Warner says is a money losing operation into a more profitable division, generated an intense wave of national protest. The protest, organized largely by the same groups and individuals that have been fighting *Channel One* ranging from Ralph Nader to Phyllis Schlafly, Consumer Reports to the Eagle Forum, succeeded in having Time Warner (then AOL-Time Warner) withdraw its plans. In many ways, the small flurry of controversy over such a potentially minor revenue flow and Time Warner's response probably had the effect of enhancing the "prestige, non-commercial" brand for the Time Warner product and helped maintain the myth that the audience can play an important role in programming decisions—the industry mantra of "we just give the people what they want."

There are two video news programs for youth produced for general distribution. *Nick News* is an occasional production of Viacom's MTV Network Services under the Nickelodeon division—sometimes producing only one new episode per month that is then repeated and repeated. Viacom also produces *Nick News Special Editions* on occasional basis and cablecasts them in the early morning for prerecording. The Weekly Reader/Time Warner *Eyewitness Kids News* is produced and aired weekly as of fall 2003. All three of these video news programs, like *Channel One*, carry conventional advertising.

The five major kids' news magazines (and matching web sites) include *The Weekly Reader* group for K-8 readers, *The Weekly Reader*/Washington Post's *Teen Newsweek* for sixth to eighth grade readers, the Scholastic Inc. group for K-8 readers, and the Scholastic Inc./New York Times joint venture Upfront for teens (presumably in the 12–18-year-old age range).

This survey could never be complete since there are a number of smaller, even harder to find sources of news for kids. To go any further would take us into terrain even further from the general public consciousness of the world of news for kids. This does not means that these sites are not important and that they do not deserve mentioning. In fact, it is just the opposite, and in the conclusion several important ones are mentioned.

Conclusion

As the 2004 election year was starting to heat up, Urban Outfitters released a new t-shirt in its "street-wise" fashion collection with the slogan "Voting is for Old People." The release raised a ruckus from a wide range of civic and activists groups working to get the youth to vote. One group, Punkvoter.com, wrote to Republican supporter and CEO Richard Hayne claiming that the effort was "a disgusting effort to reap profit from cynicism while suppressing civic involvement, and encouraging apathy." While the company complained that critics could not see the humorous irony of the t-shirt, it was pulled from their stores and online catalog.

Urban Outfitters Inc., like an increasing number of corporations marketing to youth, makes an aggressive effort to promote an antiadult world message through irony and messages of pseudorebellion. The company profile that includes its other "hot" fashion lines, "Anthropologie" and "Free People" states:

> Our established ability to understand our customers and connect with them on an emotional level is the reason for our success.
>
> The reason for this success is that our brands—Urban Outfitters, Anthropologie and Free People—are both compelling and distinct. Each brand chooses a particular customer segment, and once chosen, sets out to create sustainable points of distinction with that segment. In the retail brands we design innovative stores that resonate with the target audience; offer an eclectic mix of merchandise in which hard and soft goods are cross merchandised; and construct unique product displays that incorporate found objects into creative selling vignettes. The emphasis is on creativity. Our goal is to offer a product assortment and an environment so compelling and distinctive that the customer feels an empathetic connection to the brand and is persuaded to buy.

The corporation also received a significant amount of free press in late 2003 as the major distributor of "Ghettopoly," a Monopoly take off that outraged Black leaders, where "playas" became pimps and could draw game cards such as "You got yo whole neighborhood addicted to crack. Collect $50."

Their fall 2004 line up of t-shirts included slogans such as "You say tomato, I say fuck you" and "Everything's Dirtier in the South" and "Beer, it's what's for dinner!" and, perhaps in an effort to counter the bad publicity from its earlier *anti*-vote t-shirt, Urban Outfitters released its "I like to get it on with boys who vote" t-shirt. At the same time, third quarter earnings reports for Urban Outfitters Inc. were dazzling fashion industry investors with an 85 percent increase.

Civic education groups continue to be the primary organizations interested in the growing political apathy and political disenfranchisement of youth from our political system. While there was a slight increase in voter turnout among young voters in the 2004 presidential election, the increase turned out to mirror the proportionate increase in the overall electorate. It did not appear to be the "slacker uprising" that documentary producer and author Michael Moore campaigned for and other get-out-the-youth-vote organizations had worked toward.

When youth appear on the radar screens of U.S. corporations, it is most likely to be in the cross-hairs of their marketing departments than in corporate conversations about growing the enfranchisement of the youth vote. And this attitude appears to hold nearly as true for the news industry as well as the producers of more general products targeted toward youth. If anything, the news industry and the kids news industry appear to be more hypocritical about their intentions than mainstream manufacturers. Urban Outfitters Inc., unlike Scholastic or *Weekly Reader*, make no claims to be in the business of raising citizens. In fact they are quite pleased to be clear about selling a vague kind of rebellion against responsible adulthood and in favor of parodying adult culture and celebrating adolescent sensuality.

While studies conducted by the Kaiser Family Foundation have found that youth who follow the news are more likely to be more politically knowledgeable and have more civic engagement skills, for the most part kids, like their parents, continue to turn away, for the most part, from following the news, a trend that accelerated between the 2000 and 2004 national elections. The exceptions were cable news that has held onto a relatively constant audience of young people for news, and alternative sources such as the Internet, television news magazines, and television comedy shows that have shown dramatic gains.

The argument can be made that turning off mainstream corporate news, which has become increasingly sensational, distracting, superficial, and more in line with the corporate values of their owners and the unexamined statements of politicians, might not be a bad thing. This would depend, of course, on how that gap in knowledge about how the real politics of money and government were affecting our everyday life. Tuning into whatever shows up first on a Google search, or a television news magazine, or a television comedy show is probably not the answer.

What news should our children be seeing and hearing and discussing in schools and at home? This is an enormously difficult question because it demands of educators, parents, and civic leaders to own up to the democratic values we believe are important and that our children should be engaged with not just in the news but built into our educational priorities. Do we want to be educating young citizens in the knowledge and skills to figuratively *and literally* take government into their own hands in not only their own interest but also in the interest of the greater good of their communities, of the public? That is of course the idea of participatory democracy.

Or do we want them to learn primarily the begrudging formal responsibilities of being a citizen? Vote, do but not engage in organized dissent. Pay taxes but see this as a kind of government-sponsored robbery. Obey laws but do not raise questions about the economic or social justice of those laws. And trust that real freedom and liberty are best found in the "free" market. This is where freedom of choice takes on its truest form, where our deepest desires are expressed through the goods and services we buy or charge, without the interference of government regulation or oversight. This is, of course, the reigning neoconservative vision of democracy of our times, sometimes called corporate or free-market democracy. This is the vision of democracy that political scientist Richard Merelman contained within what he called "symbolic" civic education, where political, business, and civic leaders identify the growing decay of citizenship, but then offer only symbolic solutions to the problem, designed to make everyone feel better, but largely leave the real problem untouched.

If it is participatory democracy we want for ourselves and for our children, then we have to ask hard questions about the values built into the news they see and hear, not to mention the textbooks they study and are tested on. We have to ask questions about the lack of school funding as corporations and wealthy individuals successfully lobby federal, state, and local government to have their taxes cut, and then show up at the school doors offering new educational products and services that are supposed to help make up in efficient instruction what has been lost in funding and donations and grants with endless marketing strings attached. Is rewarding, in the midst of a youth obesity epidemic, our struggling young readers with Pizza Hut pizza parties and Nickelodeon Game Lab events what we really want? Or asking our teenagers to consume ever greater amounts of soft drinks on our school campuses in order to buy textbooks and athletic equipment?

Or do we want to be educating our children to ask the hard questions of democratic culture, like "where did the money for our schools go?" If they log onto Scholastic Incorporated's senior news site, do we want them to read only the unfiltered rhetoric of administration spokesmen and women about the triumph of the No Child Left Behind act, spearheaded by our "education" president, or also about the concern expressed by governors and educators about the failure to fully fund the legislation, leaving it a tangled web of new financial burdens on already overextended school districts and staffs?

The press is the handmaiden of democracy. As James Madison put it, "A popular government without popular information, or the means of acquiring it, is but a prologue to a farce or a tragedy, or perhaps both."

Kids need age-appropriate news—but this does not mean just news about hamsters for fifth graders or White House propaganda celebrating a mindless, unquestioning patriotism (see whitehouse.gov). Our youth need to be taught that democracy is an ever-changing idea with many faces and many names. They need to be taught about the struggles to widen and deepen the idea of democracy in the United States and to be taught that these achievements were often won through dissent and political activism, through challenging existing ideas of democracy, and imagining and fighting for a more inclusive, more equitable vision of democracy.

Returning to the central theme of this chapter, "Bad News for Kids," can mega-corporations be the willing messengers of news that empowers our children in the understanding of a "people power" vision of democracy as opposed to corporate democracy?

Parents and educators and civic leaders must get straight the version of democracy they want to promote and encourage our children to understand and embrace. Youth see news in school and are surrounded by it outside of school, whether that news comes from adult news venues like the networks' evening news broadcasts or the news contained in youth popular culture including reality video games. Games like "KumaWar: The War on Terror" game that, according to Pacific News service, "allows players to re-enact scenes of actual battles.... the game includes clips of actual video news footage, satellite pictures of Iraq and publicly available reports from the military, including the killing of U.S. soldiers" re-creating real war news events within weeks after they occur.

Once parents, educators, and civic leaders get clear about the vision of democracy they want to promote, they can begin to ask what stories about democracy, citizenship, and patriotism are being told in corporate-produced news for kids and corporate-produced learning materials and textbooks. Do we really want to be helping to sell to our children a vision of democracy that views the "public" in public education as anti-American and antidemocratic because it challenges the unregulated expansion of "free" market values to more and more aspects of our lives and our children's lives?

Bad news for kids is news that adults do not view as part of larger curriculum of civic education, at the same time defining the core values of democracy while delivering the disconnected facts of current events. More than 50 years ago, the Hutchins Commission, convened when the press was first beginning to encounter the crisis of democracy versus for-profit journalism, made a series of recommendations for the operation of the press. It articulated a code of social responsibility for the press that began with the need for the press to deliver "a truthful, comprehensive, and intelligent account of the day's events in a context which gives them meaning." Giving new urgency to this concern is the "State of the News Media" published in 2004 by the Project for Excellence in Journalism that warned "Those who would manipulate the press and public appear to be gaining leverage over the journalists who cover them."

For now, perhaps the best news sites for our youth are the web sites connected with the PBS programs mentioned above: *Frontline, The News Hour with Jim Lehrer,* and *Now with Bill Moyers.* The New York Times/Scholastic joint venture magazine "Upfront" and its companion web site are valuable and has not been overly contaminated by the Scholastic marketing machine. This is most likely due to the credibility value the New York Times needs to maintain to preserve its status as the nation's "newspaper of record." The publicly funded Canadian Broadcasting Company has excellent youth news centers for kids and teens at its web site (see [http://www.cbc.ca/]). This is also true of the publicly funded British Broadcasting Service with its *Newsround* news service for youth (see [http://news.bbc.co.uk/cbbcnews/]).

Our children must understand that democracy is not an achievement, but an unfinished journey and a struggle that pits the values of the marketplace against the values of community and collective, inclusive decision making. Our public schools must be supported to serve up the good news for kids that strengthens their commitment to a participatory democratic culture.

Chapter 9

Meritocratic Mythology: Constructing Success

Benjamin Enoma

Historically the notion of success in the United States has been largely attributed to individual responsibility and the arduous application of one's mind and physical energies to the completion of an activity or the attainment of a goal. The forgoing notwithstanding, there are factors other than individual efforts and commitment that mitigate success, and these factors are socially constructed. "One such factor is a belief that is deeply rooted in the American ideology of individualism, a belief that each individual determines his or her own situation" (Farley, 2000).

The social inscriptions of the notion of success make it susceptible to multiple interpretations. Success can be viewed as the accumulation of capital in a variety of forms: economic, social, cultural, and symbolic capital. It can be seen as a mark of distinction, the accomplishment of an individual goal measured vis-à-vis others. This measurement usually takes into account the variations in speed, quality, and mechanical accuracies involved in the completion of an activity. Lastly, success can be viewed as meritocratic, that is, the sum of individual intelligence plus effort. The common denominator in these divergent views of success is that it is always viewed or measured in relation to someone or something else.

Schooling introduces the meritocracy to most young people today. Success in school is measured by tests—standardized tests. The meritocracy is a formal system by which advancement is based upon ability and achievement. This system began with the Civil Service Reform in England in the 1870s. The idea that status should be achieved by merit, not ascribed by birth, replaced "nepotism" as the system governing the greater part of English society, and meritocracy soon transcended the European continent and quickly became established as part of the American academic tradition. Thus, the historical assumption that success is achieved by the meritocracy ideology sank its roots in the American psyche. The harder one works, the greater the chances of success. "You reap what you sow," hard work equals success to the exclusion of any salient systemic factor that works for or against some people in a given social context. Johnstone (1992) believes that in the present academic meritocracy, especially at the high and low extremes, measured merit is linked to circumstances of birth such as socioeconomic status, race, and religion. Ironically these are the circumstances that the meritocracy is supposed to stamp out. He suggests that the "university" find ways and means of selection other than mere measured merit; ways and means that better transcend the status of birth, that is, social and cultural capital (Johnstone, 1992).

Zink asked these pertinent questions in her 1997 article:

> Should not the student who has struggled to get to the doors of academe be given the same chance as the student who finds it an easy step to take? For is not education in itself an unending struggle to seek the truth and is it not in the struggle that one learns? The meritocracy sometimes sends an old, but negative message to those that struggle, saying that the doors are not open. As we near the twenty-first century, is this not a wrong and vacuous message to send, at least initially? (Zink, 1997)

When people believe that the system is fair and that the playing field is level, they usually do two things: First, they blame the unsuccessful for any disadvantages that they may experience rather than blaming the oppressive aspects of the system. Second, they oppose inter alia, policies aimed at leveling gaps (which are often conceived in terms of deficits) between the successful and the unsuccessful.

Perceptions

There are often differences of opinions in the perceptions of success by the individuals being measured and those measuring. Individual wealth is not the only indicator of success. Thomas J. Stanley in his book *The Millionaire Mind* looked at 733 self-reported millionaires. As part of his study, the author tabulated a survey that looked at 30 "*success* factors" and the results showed that "being honest with all people" was ranked as the forty-first factor. "Being well disciplined" was second, and "works harder than most people" was ranked fifth. At the other end of the list, "attending a top-rated college" was ranked twenty-third, and "graduating near the top of one's class" was last (Stanley, 2001).

In the same vein, a Gallup poll conducted on the "haves and the have-nots" and published on July 6, 1998 revealed that about one-fourth of those polled by Gallup said they were "have-nots," which is twice the rate reported as living in poverty by the federal government. More blacks than whites classify themselves as have-nots (38 percent versus 22 percent), and 43 percent of Hispanics classify themselves as such. But this disparity in self-assessment is clearest at the lower income ranges. When matched for income, blacks and Latinos with incomes of under 50,000 dollars are more likely than whites to say they are the have-nots. But blacks, Hispanics, and whites all had the same responses when their incomes were above 50,000 dollars (Gallup, 1998). One can infer from this survey that the barometer for measuring success by income is fraught with polysemous interpretations in different cross-sections of the population. Free enterprise and free market economies work under the unspoken assumption that all its citizens and/or participants will not succeed equitably. This assumption is akin to that of the social Darwinists "Survival of the fittest of the specie."

Assessment and Testing

Students' success can be defined from multiple theoretical perspectives, with each definition evoking dynamic discourses on associated concepts and facts. Consequently, students' success has to be defined "*sui generis.*" In the context of this study, it means admission into a postsecondary institution, enrollment in a program of study, and timely graduation from the course of study. The road to success in school is paved with measures and methods of assessment. Psychologists like Alfred Binet and Lewis Terman amongst others led the field of psychometrics in the development of IQ tests; these milestone tests were designed to relate the mental development of a child to the child's chronological age: $IQ = (MA/CA) * 100$. The IQ was equal to 100 times the *mental age* divided by the *chronological age*. This test is the precursor of the now popular battery of standardized tests, like the Scholastic Aptitude Test (SAT)—developed by Carl Brigham—and many more, for example, Law School Admissions Test (LSAT), Medical College Admissions Test (MCAT), Business-Schools: Graduate Management Admissions Test (GMAT), Graduate Schools: Graduate Record Examination (GRE), all of which guard the gates and entrances into most U.S. postsecondary institutions. The SAT was born from the initial IQ tests, written by French psychologist

Alfred Binet. In the United States, Lewis Terman and Robert Yerkes promoted the IQ test and made it a popular instrument to determine who should be an officer, in a segregated military, during World War I. Their IQ test was designed to prove the genetic advantage of races they had already identified as superior. Terman and Yerkes were executives in the American Eugenics Society (AES). This method of assessment is fraught with long-standing controversies such as What is intelligence? Is it a biological endowment or the product of the individual's "zone of proximal development à la Vygotsky?"

Tracking Historicity

Politics and economics have always shaped U.S. public policies. The Conservatives (traditionalists) worked to maintain the preexisting social order, which meant very remote possibilities of upward mobility or social elevation for cultures other than the dominant, and races other than the majority. The issue of racial segregation, population growth among minority groups, influx of immigrants, and the affirmation of cultural pluralism spelt new and complex challenges to a common school system. Whereas the focus at the beginning of the century was on Americanizing every school child, a shift of recognizing ethnocultural differences and a move toward multiculturalism began to emerge in the latter part of the century. Liberals and radicals (progressives) alike focused on bridging the ever-widening gap between the majority and the minority/poor students. Some progressives clamored for schools to assume the role of effectuating social justice, fostering equal opportunity for all school children and accommodating the special needs of individual children. International events (such as World War II, Sputnik, Vietnam, and the Cold War, etc.) had reverberating effects on national educational policies and citizenship. Testing and standardized testing assumed preeminence in the school system and offered a solution of sorts to the age-long controversy "Whose call is it in the school system to determine the destiny of students?" What are the bases for making that call? As early as 1892, the Committee of Ten chaired by Charles Elliot, then president of Harvard, convened to discuss uniform college entrance requirements. The Committee of Ten concluded that education for life was the same as education for college and thus denounced segregated curriculum and tracking. There were others who believed that the costs of not tracking were too great; it retarded the progress of the brighter students to wait on the slower students to catch on. It presented the teachers with the Herculean task of teaching an incongruent group of learners. Testing in the elementary schools led to segregated curricula. The high scorers were placed on the college track and the low scorer on less challenging vocational/life adjustment curriculum.

Tracking and testing are chiefly "administrative progressives" reforms. Advocates of social efficiency in education believed that since natural endowment of intelligence varied considerably across diverse groups, the best way forward is to design standard aptitude tests based on a body of knowledge that children should possess at a certain stage of development and from the results, track those who should pursue the academic or college bound curriculum and those who should opt for the life adjustment curriculum, thereby creating differentiated curriculum within an integrated system.

On a side note and contrary to popular belief, African Americans' lot in life was not ameliorated by integration. "The ink on the Brown v. Board of Education landmark decision 1954 to desegregate schools was not dry, when backlash policies such as federally aided housing programs and restricted covenants, precipitated the white flight from the inner cities in the north to the suburbs" creating in effect "de facto" segregation. Segregated housing meant, in reality, segregated schools: Citizens settled in segregated housing developments in segregated neighborhoods and de facto segregation in schools occurred as "de jure" segregation was lying in state for interment as ordered by the U.S. supreme court "with all deliberate speed."

The schools in the inner city were ill equipped and badly run, and consequently the performance levels were below predicted expectations and this unfortunately continues to date. There is an established connection between race, class, and location. Kantor and Brenzel (1992) observe that

> Urban poverty was not limited exclusively to African Americans and other people of color, but the concentration of poverty populations in central cities was strongly linked to the changing racial composition of urban areas. Poor African Americans and Hispanics were more likely than poor whites to live in low-income neighborhoods. Although nationwide and overall African Americans and immigrants populations succeeded in greater numbers than in the past, in the inner city schools, educational outcomes consistently lagged behind the outcomes in the suburbs. (Kantor and Brenzel, 1992)

Democratic Education

The underlying purpose of schooling in a democracy such as the United States is to serve as a public good helping to build sociocultural relationships between the student and his or her community. This charge presupposes that there is no discrimination and that there is equality of access and unfettered ascent through the classes in the school system. The reality, however, is that schools are as fragmented as the society at large; every learner arrives wearing his or her "Habitus": social and cultural capital. With reference to capital, mainstream students characteristically bear a comparative advantage over minority students, a headstart so to speak. This disparity notwithstanding, both groups are measured for success by the same battery of standardized tests, along age groups and class levels.

Predictably there emerges noticeable achievement gaps stratified along racial and socioeconomic lines; the purpose of schooling in this democracy is thus compelled to respond in the face of these glaring inequalities. How does egalitarianism coexist with the meritocracy? Who will bear the costs of reforming this public good?

The responses to this matter are steeped in politics and power relations and the fault lines are drawn along epistemological and ontological positions. The conservatives, the liberals, and the radicals all etch their arguments for school reforms from their political convictions. The conservative viewpoint is the position that seeks to preserve the dominant culture or the established order. Attempts to revamp the status quo pose a threat to the conservative position. The liberal position is accepting of the dominant culture yet broad minded; in that it is also open to reform or propositions for change. Finally, the radical position is often critical of or dissatisfied with the status quo. This viewpoint proposes alternate solutions and denounces the conservative viewpoint. The radical is often touted as an extremist standpoint because of its preoccupation with denouncing the established order. All of these positions represent separate ideologies, informed by different logics and schools of thought.

The struggle with school reforms has not only focused on testing and tracking but also on the very purpose of schooling. A watershed event in the arena of curricular reforms is John Dewey's significant, much-maligned, and often misinterpreted treatise: *The School and Society*. This work was pivotal on many fronts. First, Dewey was advocating the higher ideal of democratic education; since the "child's life is an integral, total one," the school should reflect the completeness and unity of the child's own world. Dewey advocated active learning, starting with the needs and interests of the child; he emphasized the role of experience in education and introduced the notion of teacher as a facilitator of learning as opposed to the "font from whom all knowledge flowed" (Sadovnik and Semel, 1998). Enter the progressives, "Learning by doing." Child centered progressives spun off Dewey's treatise and wanted to achieve the lofty goal of democratizing education but in implementation, progressive education (pedagogical progressives) became feverish pursuits of the elusive democratic education. Progressive education in practice became essentially democratic education for the elite.

Under the current framework, there is still a search for the common school, uniform curriculum, uniform assessment, standardized testing, and the like. What should be taught, at what level, and to whom? The social melorists and other progressives argued whether it was more democratic to teach all students the same subjects or to tailor curriculum to individuals.

Tyack (1974) on the subject of the curriculum reforms proffers a radical point of view that I agree with. Talks about keeping the school out of politics have often served to obscure the actual alignment of power and patterns of privilege. "He declares that in sum the search for the 'One best system' in the area of curriculum and instruction has ill-served the pluralistic character of the American society. Americans have often perpetuated social injustice by blaming the victim, particularly in the case of institutionalized racism" (Tyack, 1974).

These reforms have also made their mark on the administration of schools, moving them from the traditional organization management styles to problem-solving, corporate efficiency models. The "social efficiency group" emerges as the expert here, making the case for running schools as businesses for improved efficacies. While there is a good case for waste and mismanagement of funds in school administration, this approach reduces the public good dimension of the purpose of schooling to a line item budget that will be approved or discarded based on costs–benefits analysis.

Critical Discourses

Student success can be looked at panoramically as a sociocultural construct; Pierre Bourdieu in the *Forms of Capital* expands capital beyond its economic perspective, which accentuates material exchanges, to include noneconomic and immaterial forms of capital. He favors a nurture-rather-than-nature argument throughout his discussion on the forms of capital. He argues that an individual's talent and ability are primarily determined by the time and cultural capital invested in them by their parents. Similarly, he posits than an individual's scholastic yield from educational action depends on previously invested cultural capital by the family. One can infer from this as in his notion of habitus that students' success can be regulated, reproduced, and predestined with the investments in cultural capital made by the families of the students in question. Cultural capital exists in three forms: (1) embodied within the individual; (2) objectified in artifacts and resources, which can be appropriated materially via economic capital or symbolically via the embodied capital; and (3) the institutionalized form that refers to various instruments of legitimization, academic credentials, and professional licenses. In this viewpoint, the die is not necessarily cast in favor of the wealthy. Although it helps indubitably to possess material and economic means, access to cultural capital can be made through noneconomic means such as social networks.

Success can also be viewed dichotomously via the subparts of academic achievement and on the job performance or formal academic knowledge and practical intelligence (tacit knowledge). Sternberg et al. (1995) argue that practical intelligence and tacit knowledge parallel academic intelligence and formal academic intelligence, respectively. An academically intelligent person is deemed to be so because he or she has acquired formal academic knowledge and has been tested through a wide range of intelligence and aptitude tests. By contrast, the practical intelligent person has acquired tacit knowledge that has been tested through various real-world events but is not predicted through conventional intelligence testing.

As Michael Polanyi (1967, p. 4), a precursor in this field, posits in his work, *Tacit Dimension*, we should start from the fact that "we can know more than we can tell." He termed this prelogical phase of knowing as tacit knowledge: it comprises a range of conceptual and sensory information and images that can be brought to bear in an attempt to make meaning. Although importance is ascribed to this notion of tacit knowledge, the voices of dissent should also be acknowledged. Jensen (1993) asserted, "tacit knowledge seems an exceedingly mysterious variable, theoretically and empirically." According to Jensen, it is neither a personality measure nor a predictor of scholastic performance. More empirical support is needed on how best to measure tacit knowledge for the concept to become theoretically grounded.

Somesh and Bogler (1999) scrutinized the main themes of tacit knowledge; "informal and implicit knowledge used to achieve one's goals." Undergraduate students were scrutinized: students' socioeconomic status (SES) and gender were also examined for variance in tacit knowledge and how it relates to academic achievement. Employing a questionnaire consisting of biographical information and a tacit knowledge scale that they developed, they found that students with low SES made more use of tacit knowledge than students with high SES. This finding, if validated, is particularly interesting to my study because low SES is usually associated in a lot of studies with deficiency; here, however, what is usually referred to as "street smart" or tacit knowledge is appropriated more by students with low SES. Tacit knowledge is not an unconscious possession but a subconscious, latent ability that is appropriated on the basis of needs.

What difference would it make if tacit knowledge were valued in the admission process? What if ways were explored of assessing different discursive knowledges in the admissions decisions? Furthermore in the study, it was revealed that students with high tacit knowledge achieved higher academic grades than students who had low tacit knowledge. The significant inference drawn here concerns the importance of tacit knowledge to student success in learning institutions. The nature of tacit knowledge is intrinsic to the individual, that is, action-oriented knowledge acquired without direction or help from others, yet it allows individuals to appropriate resources to achieve the goals that they personally value. They are able to utilize resources like the reference library, tutorial services, and academic advisors and discover the "hidden curriculum."

An area of study that will be of great interest will be to evaluate factors that surround the acquisition of tacit knowledge. Is the acquisition propelled by hardship and the instinct to survive? The answer will shed some light on why low SES students rank higher in tacit knowledge than high SES students.

On the other hand, meritocracy is a formal system by which social advancement or attainment is based upon cognitive ability and achievement, which began with the Civil Service Reform in England in the 1870s. When Michael Young, the British sociologist, coined the word meritocracy in his celebrated essay, "The Rise of the Meritocracy" in 1958, it had a negative connotation. In this futuristic exposition, the social place or status of the individual would be determined by IQ plus effort ($I + E = M$). There are many criticisms of the notion of meritocracy akin to Young's original position. Some critical theorists posit that the power elite subscribe to merit simply to legitimize a system where social position is really determined by class, birth, and wealth. I agree with the foregoing argument to the extent that the pursuits of and reliance on merit in a society like the United States, where race is inextricably linked to class and social standing serves to reify the "status quo."

Seligman (1994) draws a comparison between Young's 2034 futuristic setting to Richard Herrnstein and Charles Murray's (1994) "*The Bell Curve*," praising the prescient Young for painting a realistic picture of what life has become in present-day America. Is cognitive ability central to social mobility and economic success? Robert Hauser offers a response of sorts in his working paper: "Meritocracy, Cognitive Ability, and the Sources of Occupational Success." Much of the standard psychometric evidence is weak, but ability does play a significant role in social stratification, primarily by way of its influence on schooling. There is no clear evidence of trend in the role of cognitive ability in the stratification process, and other social psychological variables may be equally important. There is no evidence that cognitive ability is *the* central variable in the process of stratification, but there is ample reason for concern that recent and prospective changes in the structure of American education will raise its importance (Hauser, 2002). Much of the controversy over meritocracy and it relationship to success lies on the issue of fairness and equal opportunity. Employing cognitive ability as the launch pad for merit is historically situated in the Ivory Tower's social engineering efforts.

James Bryant Conant, who on becoming the president of Harvard in 1933 encouraged one of his deans, Henry Chauncey, to embark on an ambitious program of educational testing are two looming figures in this account. The goal was lofty, that is, the future of American democracy crucially depended on opening up its elite educational institutions to a much wider constituency than the rich and the famous. The aim was to create, as Lemann (1999) puts it, "a scientized social utopia" by applying a standard gauge to people.

There is a body of literature showing that IQ tests are not culture free or guarantors of educational attainment. Thus, the lofty goals of Conant and Chauncey were quickly sabotaged by the aristocracy and the meritocratic recruits; they had learned how to play by the new rules and found ways to get their children into the best universities via preparatory programs like "crammers" in New York that coached students on how to excel at the college boards. Thus, the original drive of moral education and inclusion of the poor and minority took a back seat to the search for merit, which in this experiment meant total reliance on academic testing ability.

Cognition Discourses

The notion that intelligence is not an intrinsic quality that a human being possesses rather than a social and material distribution that people accomplish via interaction and activity is quite revolutionary. This claim takes proponents of eugenics and social Darwinists to task on the issue of intrinsic, individual, intellectual endowment and gene superiority. Pea (1993) posits that cognition is something a human does and not the acts of designed objects. His point delineates between solitary intelligence and distributed intelligence. The former is the largely held view in the field of education. It invokes intelligence as a property of the individual mind, its capacity to absorb new information, retain, recall, and apply its store of information in various circumstances. The humanists basically see the mind as muscle to be furnished, flexed, and disciplined through a systematic study of stru tured curriculum. This is what standardized testing and school assessments are based on.

Pea's premise is sound in that the mind rarely works alone when doing or practicing cognition. It always works in conjunction and consortium with artifacts, persons, and environments, natural or artificial. Meaning making is a confluence of activity enabled by the individual agent and the surrounding resources. How individuals appropriate the resources in the environment is of considerable interest. The notion that an individual is unable to learn because he or she is mentally weak by default of race or social class finds a new explanation. The presence or absence, and the appropriation or the lack thereof, of the "cultural capital" and "artifactual" resources necessary to accomplish intelligence in their respective environments or activities becomes pertinent.

Summation

The relationship between student success and the meritocracy is profoundly ironical; the pretext of the meritocracy is to produce a classless society through the use of competitive standardized tests in schools. The idea that "meritocrats" (high scorers) in the tests would emerge from various strata of the society, thereby redistributing social order and expanding the patterns of privilege has turned on its ear. The reality is that meritocracy over time degenerates into aristocracy. The emerging high scorers and every one in position of privilege begin to use their power and sophistication to turn the odds in their favor and assure certain rites of passage for their forebears; the so-called legacy candidates in postsecondary institutions applicant pool.

The essential flaw in the meritocracy is that it is not altruistic; meritocrats or the successful students assume the responsibility of paving the way of easier access for their kith and kin by investing their newly acquired social and cultural capital. This responsibility ultimately circumvents the range and strata of society where high scorers come from. They create, in Vygotskian terms, a robust zone of proximal development (ZPD), where the vocabulary of the tests and preparatory regimen are appropriated around the dinner tables or in exclusive prep programs.

Student success is not the exclusive preserve of the privileged and the wealthy; the odds are stacked in their favor, the privileged hold a comparative advantage over the commoners. Sadly in this meritocracy, every time a commoner crosses the great divide and succeeds, he or she by default morphs into a privileged person and may become the ferry across the gulf that separates his or her poor from the privileged.

On the surface there seems to be little that is wrong with this picture because assuredly there is a chance every once in a while that the poor can get over this great divide and in time ferry some or all of their own across, except that the privileged are who they are by the dimension of control. The privilege control access, control the tests, control the funding, control the curriculum and the administration, and have a huge say in the morphing or transformation process.

Have we essentially crossed the Rubicon in this matter? Or is there any hope for a change? My response is an emphatic yes to hope. Critical theorists underscore the importance of agency; individuals can liberate themselves from the oppression of power and shape and assume control over their own lives; the answer lies in a revolutionary epistemology and ontology of knowledge production. Kincheloe speaks of this leitmotif for critical teachers and the allure of Freire's impassioned spirit and radical love, "Critical pedagogy is an ambitious entity that seeks nothing less than a form of educational adventurism that takes us where nobody's gone before" (Kincheloe, 2004). Reeducating the world in tandem, in groups, or at the very least one mind at a time is progressive.

References

Farley, J. (2000). *Majority–Minority Relations*, 4th Edn. Upper Saddle River, NJ: Prentice Hall.

The Gallup Poll (1998). *Have and Have-Nots: Perceptions of Fairness and Opportunity*. A study based on a survey of 5,001 American adults, conducted between April 23 and May 31, 1998.

Hauser, R. M. (2002). *Meritocracy, Cognitive Ability, and the Sources of Occupational Success*. Center for Demography and Ecology, Department of Sociology, Madison: University of Wisconsin.

Hernstein, R. and C. Murray (1994). *The Bell Curve: Intelligence and Class Structure in American Life*. New York: The Free Press.

Jensen, A. R. (1993). Test Validity: g versus "tacit Knowledge." *Current Directions in Psychological Science*, **2**, 1, pp. 9–10.

Johnstone, D. B. (1992). The University, Democracy, and the Challenge to Meritocracy. *Interchange*, **23**, 1–2, pp. 19–23.

Kantor, H. and B. Brenzel (1992). Urban Education and the "Truly Disadvantaged": The Historical Roots of the Contemporary Crisis 1945–1990. *Teachers College Record*, **94**, 2, pp. 278–314.

Kincheloe, J. L. (2004) *Critical Pedagogy Primer*. New York: Peter Lang.

Pea, R. D. (1993). Practices of Distributed Intelligence and Designs for Education. In G. Salomon (ed.), *Distributed Cognitions*, pp. 47–87. New York: Cambridge University Press.

Polanyi, M. (1967). *The Tacit Dimension*. London: Routledge and Kegan Paul.

Sadovnik, A. R. and S. F. Semel (1998). *"Schools of Tomorrow," Schools of Today: What Happened to Progressive Education*. New York: Lang.

Seligman, D. (1994). Foretelling the Bell Curve. The Rise of the Meritocracy. *National Review*, **46**, 12, p. 55.

Somesh, A. and R. Bogler (1999). Tacit Knowledge in Academia; Its Effect on Student Learning and Achievement. *Journal of Psychology*, **133**, 6, pp. 605–617.

Stanley, T. J. (2001). *The Millionaire Mind*, pp. 31–35. Kansas, MO: McMeel Publishers.

Sternberg, R. J., R. K. Wagner, W. M. Williams and J. A. Horvath (1995). Testing Common Sense. *American Psychologist*, **50**, 11, pp. 912–927.

Tyack, David B. (1974). *The One Best System: A History of American Urban Education*, pp. 11, 180. Cambridge: Harvard University Press.

Zink, L. (Fall 1997). Is the Meritocracy Necessary Even at the Doors of Academe? *Journal of College Admission* **157**, pp. 22–29.

Chapter 10

What Is Not Known about Genius

Ray McDermott

Lawgivers, statesmen, religious leaders, discoverers, inventors, therefore only seem to shape civilization. The deep-seated, blind, and intricate forces that shape culture, also mold the so-called creative leaders of society as essentially as they mold the mass of humanity. Progress . . . is something that makes itself. We do not make it.
Alfred Louis Kroeber,
Configurations of Culture Growth, 1944

As for attributing "genius" to men who have "changed the course of history," we have seen that an idiot or a goose can accomplish it just as well. It is not high or low levels of ability that is significant in such contexts; it is being strategically situated in a moving constellation of events.
Leslie A. White, *The Science of Culture*, 1949

Contrary to the anthropological opinions of Kroeber and White during the mid-century, genius is taken by current practice to be a kind of person with extraordinary intelligence who gets used to accomplish great things. If the difference between a normal person and a talented one is that of degree, then the difference between a talented person and a genius is that of kind. By this logic, Solieri was talented and Mozart a genius. New generations define genius by matching seeming solutions to apparent problems, and exemplars of genius—a new Newton, Balzac, Joyce, or Einstein—is identified, explained, and celebrated. Every generation also worries about lost possible geniuses. Sometimes a forgotten or ignored genius is remembered. Only rarely is the very idea of genius confronted. The genius drama needs a new analysis.

The first section of this chapter offers a skeleton history of the term genius from 1650 to 1900 and identifies a transition from a renaissance genius as a momentary medium for inspiration to a particular kind of special person with an established position in society. By 1750, says Kenneth Frieden, "a craze of theoretical writings urges that the inspired need not *have a genius*; instead, the inspired author *has genius* or *is a genius*" (1985, p. 66; emphasis added). By 1750, the genius was a position in European society. Once the throne was built, it had to be filled. When there were no geniuses available, the throne had to be filled nonetheless; if there were a plethora of geniuses available, one person had to be put on the throne nonethemore. As the category genius was ill defined, the attribution of genius was an opportunity for political mischief. A genius should be more intelligent than less, yes, and more accomplished than less, yes again, but these slippery words had a direction: The genius list was composed of white males from powerful European states. Whatever the categories, however applied, the results were exclusive. By 1750, genius was located only in an occasional, white, male, individual mind. Genius thereafter has been often in the service of unjust hierarchy. Inherent genius has been conducting business—literally business—in a niche that delivers a hierarchy of the smart over the seemingly less inventive and less creative.

The second section of this chapter gathers confrontations with genius as a kind of person with high intelligence. Centuries of counters easily expose the political mayhem invited by conceptions of genius as an inherently gifted and worthy person. The chapter also critiques institutionalized uses of *inherent intelligence* as a category that mocks the complex lives of children. Anthropologists have tried to disrupt learning categories that acquire people near the bottom of the social order—the illiterate, the inarticulate, the learning disabled (e.g., McDermott, 1988, 1993; Varenne and McDermott, 1998). This chapter does the same for a kind of person in the luxury seats. Confrontations with genius between 1650 and 1900 are resources for current struggles—Kroeber, White, and then some—against the easy use of inherent ability as an apology for inequalities in opportunity and achievement across gender, race, and class borders. Named positions in a hierarchy usually have more precise borders—arbitrary, yes, but replicably so—than the complex persons they acquire or exclude. The precise mismeasure of intelligence has been institutionalized in schools and popular media as a tool for ignoring the complexity of children. Genius has the top position in a hierarchy of intelligence types that supports a theory of the mind often used to keep everyone in place.

Section three points to ties between the attribution of intelligence and already established inequalities. The genius figure has been used to erase women, racial minorities, and school children of all kinds. Name ten women geniuses. It is possible, but only after research and argument. Now name ten recognized African American geniuses. The next one seems easier: Name ten Jewish geniuses. Now it gets difficult: Name ten Jewish geniuses in Europe in the 1930s and check their fate in the 1940s. Whether in naming or maiming, there are inequalities in genius sighting. Inherent genius might be a bad idea with bad consequences. As an invitation to justify and/or degrade the cognitive best over the rest, naming a genius can be an occasion of social violence.

The news offered in this chapter is old—and unfortunately forgotten. For the past 50 years, anthropologists have been writing critiques of what happens to children in American schools, and many have critiqued psychology for supplying an easy mentalistic language for describing, diagnosing, and explaining differential ability and intelligence. Cognitive and psychometric psychology have become so institutionalized in commonsense that they are part of what any ethnographer interested in thinking and learning must reject. How nice it is to find that genius—a key category in psychological accounts of what makes people special—has been held in suspicion and dramatically critiqued for hundreds of years before current biases took root. This chapter tries to recover the critical tradition.

Genius, Position Of

In 1711, Joseph Addison[1] announced a new object for public scrutiny: those few who "draw the Admiration of all the World upon them, and stand up as the Prodigies of Mankind, who by mere Strength of natural Parts, and without any assistance of Art or Learning, have produced Works that were the Delight of their own Times and the Wonder of Prosperity" (p. 282). He listed first Old Testament poets, Homer, and Virgil. Next came Pindar and Shakespeare. There were no scientists on his list, although Newton's genius was in the air, what with him standing rarified on the shoulders of others (Koyré, 1952). Everyone had an opinion on genius. Along with an Enlightenment commitment to reason, Denis Diderot, from 1750 on, celebrated the imagination of individual genius. Herbert Dieckmann makes the strong case: "The transition from the conception of genius as mere talent to the conception of the genius as an individual was accomplished through a specific act of thought... Diderot accomplished this act of thought" (1941, p. 152; see also Jaffe, 1980). For Diderot, even government must cater to the inspired—encourage them, yes, and put up with their foibles. If forced to choose between Racine, a great person without poetry, and Racine, a jerk with poetry, society must settle for and celebrate the poet—"De Racine méchant que reste-t-il? Rien. De Racine homme de genie? L'ouvrage est éternal" (cited in Dieckmann, 1941, pp. 181–182). There must be a place for the eternal work of genius. Racine "was of use only to people he didn't know, at a time when he ceased to live" (Diderot, 1964, p. 15). Diderot's point is embodied by Romanticism for the next 50 years, and the genius becomes not only a kind of person who/that exists, but also one that/who is easy to see. By 1832, Honoré de Balzac reports that "to most biographers the head of a man of genius rises above the herd as some noble plant in the fields attracts the eye of a botanist in its splendor" (2002, p. 2). Getting seen by the right people became the only problem. Genius had become a goal, something to strive for, to be discovered as—ahhh, to be seen rising "above the herd."

Through the twentieth century, framing genius as a kind of person became unremarkable. It is now commonsense. Popular movies and plays feast on the tensions of being smart, smarter, and smartest (e.g., *A Beautiful Mind, Searching for Bobby Fischer, Good Will Hunting, Finding Forrester, Genius, Proof*); newspapers print top-ten genius lists; hundreds of books advise how to unlock individual genius; hundreds more celebrate a specific genius; psychologists measure both genius and the potential for genius; social scientists explain individual genius, why this person and not that, why more from one group than another; university admissions seek potential geniuses; and genius clubs sponsor shared activities from math problems and puzzles to dating.

How did the shift take place, and why does it now make easy commonsense to talk about genius in individualistic terms? One answer comes from seeing genius in the context of co-occurring political and economic events: the rise of capitalism, colonialism, democracy, racism, and individualism. Struggles with genius by three thinkers a century apart from each other illustrate the changing contexts of their ideas. Blaise Pascal (1623–1662), Adam Smith (1723–1790), and Francis Galton (1822–1911) make a continuum on genius from ill-placed praise for a man of science in the service of God (Pascal), through irrelevant praise for a successful participant in a reasonable and productive economy (Smith), to rightful praise for a person naturally gifted with a propensity for great achievement (Galton). Two hundred years of radical economic change after Pascal, Galton marks the start of a commodified genius: quantified intelligence on sale as a product that promises breakthroughs and still more production. For Galton, Pascal's God becomes nature's loaded dice box that keeps coming up with great intelligence for English males, and Smith's rational economy is made into a display board for the grandeur of great minds, operationalized by Galton as "men of eminence." As Pascal is shockingly and Smith surprisingly different from the present mindset, I offer Pascal, then Smith, and finally Galton as contrasting moves along the way to current biases.

A comment on sources: Histories of genius and related ideas do not typically feature Pascal's resistance or Smith's dismissal and are limited to more enthusiastic theories of artistic and scientific genius (Nahm, 1965; Engell, 1981; Murray, 1989). A usual lineup starts with Addison in 1711 and continues with a French Enlightenment figure (say, Diderot), a Scottish Enlightenment celebration of genius (Duff, Gerard), Kant, a Romantic vision (early Goethe, Coleridge), often Thomas Carlyle, and finally a Darwinian and Freudian theory. This chapter's list—Pascal, Smith, and Galton in this section, Henry Fielding, Ralph Waldo Emerson, and George Plekhanov in the next—roughly follows the same outline, but with more complaints about the injustices of genius touting. To Addison and Diderot, I have added the earlier views of Pascal, Smith represents the Scottish Enlightenment, Emerson is the more profound partner in dialog with Carlyle, Galton develops a Darwinian thesis, and in Plekhanov, I offer not a Freudian, but a Marxist critique of the hero and genius figure. Together these sources offer forgotten critiques from a wide range of disciplines (science, philosophy, economics, literature, and psychology) and nationalities (English, French, Scottish, Russian, and American; the last rarely figures in mainstream histories of genius theory).

In 1652, Pascal was becoming the toast of Europe for his scientific work on the geometry of conic sections and the vacuum (and more to come with early versions of calculus and probability theory). After he sent his new calculating machine to Queen Christina of Sweden, she said she admired Pascal for his clarity and commitment to proof over appearances. A friend reported that the Queen thought Pascal was "one of those geniuses for whom the Queen had been searching" (Cailliet, 1961, p. 96). Pascal was taken aback. He had two public loves, science and God—or in a more secular or operational vein, inquiry and humility—and he felt a conflict between them. Knowledge and fame were both the means and end of science, and the humble acknowledgement of ignorance performed the same dual service for his theology. To be "one of those geniuses for whom the Queen had been searching" is difficult to do with humility? Let us call this Pascal's problem: As an invitation to self-pride, being called a genius is an occasion of sin.

Pascal tried to give up science to save his humility, but soon found that he was so proud of being humble, he might as well be a practicing genius. He sounded a warning:

> Discourses on humility give occasion of pride to conceited persons and of humility to the humble. Even so, those on skepticism lead dogmatists to dogmatize. Few men speak humbly of humility, chastely of chastity, or skeptically of skepticism. (Fragment 255; I cite the *Pensées* of 1670 by Fragment number only)

A healthy tie between knowledge and humility is difficult to maintain in the face of adoration. Pascal said little about genius—only that great intellect is much better than power and riches, but infinitely less worthy than charity (Fragment 852)—but he did offer a social theory of authorship:

> Certain authors, speaking of their works, say: "My book," "My commentary," "My history," etc. They resemble middle-class people who have a house of their own and always have "My house" on their tongue. They would do better to say: "Our book," "Our commentary," "Our history," etc., because there is in them usually more of other people's than their own. (Fragment 43)

The assault of pride on humility is softened if others are acknowledged, although award ceremonies show how individually royal a "we" or an "our" can be. Humility that brings pride to the proud reveals a deeper problem. For Pascal, "we are nothing but falsehood, duplicity, contradiction: we both conceal and disguise ourselves from ourselves" (Fragment 255). Neither rejecting science nor embracing humility could release him from bad faith. Genius and humility are not paired opposites, but each is an occasion for defining and contrasting the other in an unending sequence of concealment and disguise. Pascal used grandeur and wretchedness to illustrate his case:

> Wretchedness being deduced from grandeur, and grandeur from wretchedness, some have inferred man's wretchedness all the more because they have taken his grandeur as proof thereof, while others have deduced his grandeur with all the more force because they have inferred it from his very wretchedness. All that the one side has been able to say in proof of his grandeur has only served the other side as an argument of his wretchedness, for the greater a man's fall, the more wretched he is; and vice-versa. The one side tends toward the other in an endless circle, certain as it is that in proportion as men are enlightened they discover both grandeur and wretchedness in man. (Fragment 237)

Ditto Pascal on genius and humility, or science and ignorance: "The one side tends toward the other in an endless circle." Here is a translation from the English of grandeur to the English of genius:

> *Stupidity* being deduced from *genius*, and *genius* from *stupidity*, some have inferred man's *stupidity* all the more because they have taken his *genius* as proof thereof, while others have deduced his *genius* with all the more force because they have inferred it from his very *stupidity*. All that the one side has been able to say in proof of his *genius* has only served the other side as an argument of his *stupidity*, for the greater a man's fall, the *dumber* he is; and vice-versa. The one side tends toward the other in an endless circle, certain as it is that in proportion as men are enlightened they discover both *genius and stupidity* in man. (adapted from Fragment 237; emphasis added)

It is a mistake to freeze either grandeur or wretchedness into place without regard for how they together infer, require, and even create each other. The same is true for genius and stupidity. How could Pascal be a man of knowledge when the very label depended on a denial of his ignorance? How could he be a man of knowledge when the very label implied that a gift from God was his possession? How could he be a genius without confusing something borrowed with something owned? No wonder two recent books (Rogers, 1999; Bourdieu, 2000) have joined an older one (Goldmann, 1964) claiming in Pascal a social theory celebrating people on the bottom and confronting those on the top.

The term genius does not acknowledge the reticular ties between grandeur and ignorance, not in 1652 and much less today. For almost 2,000 years, the term had a stable history in Latin as a guardian spirit. The renaissance genius took a second referent as a person who, momentarily inspired by God, could perform a great task (Screech, 1983). Pascal could live with either definition, but by 1650, genius was developing the personal property meaning that would become commonsense a century later and lead cumulatively to its current sense as a person with a naturally brilliant intelligence. Humility was absented as personal pride became valued. Ignorance was hidden as pure intelligence became valued. The ascription of genius shifted from moments and their minds to minds and their moments.

That Pascal doubted the genius that heads of state might seek was not apparent in his letter of appreciation to Queen Christina. From one scientist to another, from one with no formal power to a royal leader, Pascal invoked a world that has existed for only moments in the centuries since:

> It is Your Majesty, Madame, who furnishes to the universe this unique example of which it was lacking. It is You in whom power is dispensed

by the life of knowledge, and knowledge exalted by the luster of authority. (Pascal, 1989a, p. 31)

What a good idea: Science and authority working together, bringing about a more lustrous world. In the kingdom of the smart, the large-brained man should be king. We should have a democracy of the mind. Make that a monarchy for Pascal, but he was formulating a coming world that would claim to have a democracy of intelligence based on merit and achievement. Even as he wrote to the queen, Pascal likely doubted his own dream (as he would doubt the grounds of any self-celebration). If all achievements are borrowed from the moment, if they are gifts from God, how can they be a source of a moral order run by individuals arranged in a hierarchy of ascribed genius? Are the light of science and the luster of authority to shine only through special individuals who are mentally fast, complex, articulate, and demonstrably correct? We should all have, says Pascal, "a double conception . . . if you act outwardly with men in accordance to your rank, you ought to recognize, by a more hidden, but more genuine conception, that you have nothing naturally that is superior to them. If the public conception elevates you in a perfect equality with all mankind, let the other humble you and keep you in a perfect equality with all mankind; for that is your natural state" (1989b, p. 74). Many contemporary societies are placing their bets on a version of nature that delivers an inherent best, and measures of intelligence and aptitude—academic degrees along with IQ, SAT, GRE scores—now dominate access to positions of eminence. Are minds—the modern kinds of minds that process information and solve problems—to be the units of social order and responsibility?

A century after Pascal, Adam Smith gave a second reason to eschew a kingdom of the smart:

> Moralists exhort us to charity and compassion. They warn us against the fascination of greatness. This fascination, indeed, is so powerful, that the rich and the great are too often preferred to the wise and the virtuous. Nature has widely judged that the distinction of ranks, the peace and order of society, would rest more securely upon the plain and palpable difference of birth and fortune, than upon the invisible and uncertain difference of wisdom and virtue. The undistinguishing eyes of the great mob of mankind can well enough perceive the former; it is with difficulty that the nice discernment of the wise and the virtuous can sometimes distinguish the latter. In the order of all those recommendations, the benevolent wisdom of nature is equally evident. (1976, p. 226)

For Pascal, reliance on wisdom and virtue leads to arrogance. For Smith, it leads to misperception and mayhem. Smith was not a fan of the "nature" that delivers rank by birth and fortune. He preferred market arrangements that could deliver prosperity to all. A new order would develop with the new economy, not by a celebration of individual creativity or wisdom. Better to argue for a hierarchy based on birth and wealth than to rely on the illusive subtlety and finesse required of a hierarchy based on creativity.

If capitalism is to work full steam ahead, self-interest, ambition, and even greed must supply the motivational impetus. For Smith, the raw energy of self-interest distributed across a population can lead to increased prosperity. Smith also stated limits. Contemporary right-wing apologists stress Smith's objections to state intervention in free trade, but he had a wider range of concerns. He hated monopolies most of all, and, yes, for the ways they interrupted free trade. He hated poverty too, and his texts consistently resolved conflicts between the freedoms of the rich and the suffering of the poor in favor of helping the poor (Rothschild, 2001). In a proper economy, everyone, including the poor, moves forward. What then of genius? For Smith, the eighteenth-century genius exists only as a consequence of training and, more importantly, of training delivered in a division of labor:

> The difference in natural talents in different men is, in reality, much less than we are aware of; and the very different genius which appears to distinguish men of different professions, when grown up to maturity, is not upon many occasions so much the cause as the effect of the division of labor . . . the most dissimilar geniuses are of use to each other; the different produces of their respective talents, by the general disposition to truck, barter, and exchange, being brought, as it were, into a common stock, where every man may purchase whatever part of the produce of other men's talents he has occasion for. (Smith, 1988, pp. 14–15)

In a proper economy, there would be no need for genius. Everyone's labor, "by the general disposition to truck, barter, and exchange," would be "brought, as it were, into a common stock," and so too would everyone's mental labor. "In great social questions," said Bernard Shaw much later, "we are dealing with the abilities of ordinary citizens; that is, the abilities we can depend on everyone except invalids and idiots possessing, and not with what one man or woman in ten thousand can do" (1928, p. 172). The genius is a monopoly of intelligence that holds back full participation in the market of ideas, just as industrial monopoly limits participation in the market of goods. The result, in both cases, is an improper division of labor. The poor must be educated, said Smith, in order to participate, and the smart state foots the bill (1994). The more people know, the greater the possibility labor will expand production and general prosperity.

Without contrasting Pascal and Smith,[2] we can use them as a viewing platform for what we have since produced: a kingdom of the seemingly smart, located arbitrarily, and serving a hierarchy of measured intelligence in ways consistent with and legitimizing the same class and race divisions once brought about, only more obviously, by wealth, color, and fortune. To Pascal's horror, we have advanced ties between knowledge and arrogance. To Smith's horror, we have advanced ties between success and an arbitrary categorization of wisdom and virtue. Worse than arrogance or misjudgment, we have combined them in the service of elites. We have allowed school and test results to stand for wisdom and virtue. The measure of the new person is hard cold scores and the currency that follows in their wake, and what a wake it has been: arbitrary demoralizing facts, legally binding, politically blinding. The term genius has been part of this trend. In Pascal's terms, genius has become an occasion of sin.

The term genius did not change on its own, but with a set of related terms: *creativity, individual, imagination, progress, knowledge, science, insanity, race, intelligence,* and *human nature*. Together they pointed to an emerging theory of mind well fitted to the emerging capitalism that has been the context for the institutionalization of genius. An advanced capitalist state does not have to develop an exclusively cognitive, individualistic account of the mind, but it often does. When everyone in a society is pitted against everyone else in the pursuit of riches, an apparatus that measures every intelligence against every other in arbitrary ways that correlate with already existing class and race borders is likely a growth industry. The genius under capitalism has an efficient mind: less has to go in, more comes out, the less to be worried about, the more to be sold. It is a capitalist's dream factory transferred to the mind.

In 1869, Galton introduced inherent genius in men of great eminence while controlling for education. He wanted to show that psychological characteristics worked by the same rules of inheritance as physical traits, that "out of two varieties of any race of animal who are equally endowed in other respects, the most intelligent variety is sure to prevail in the battle of life" (Galton, 1962, p. 292). He was an inventive researcher, his methods still influential. For genius, he gathered achievement data on the adopted sons of Catholic popes and bishops and found that the young men, despite an abundance of education and good connections, did not achieve anything like their mentors. He used related techniques to show the inefficacy of prayer

(Galton, 1872), but these results, controversial at the time, have dropped comparatively from view. (He did not test every possibility: If celibacy were as ineffective as prayer, the adopted sons of bishops might have had a full line inheritance from their mentors.) He presided over a marriage of Adam Smith and Charles Darwin, of self-interest and natural selection, together in an Inquiry into the Nature and Causes of the Wealth of Creations among upper class Englishmen. Eminence was male, muscular, and moneyed. Eminence was a top-draw power broker and cutthroat capitalist. Of the 13 kinds of genius he investigated, most (Judges, Statesmen, Peerages, Commanders, Literary Men, Men of Science, Divines, Senior Classics at Cambridge, Oarsmen, and Wrestlers) were categorically limited to his classmates (although he listed 4 women among his 47 Literary Men), and only the Poet, Musician, and Painter categories could easily admit women or persons of lower birth. Where Pascal feared arrogance and Smith feared the celebration of irrelevant knowledge, Galton showed no fear at all. Arbitrary hierarchy was not forsaken, but celebrated, and arrogantly so. Nature made upper class Englishmen this way. They own it. Genius is their property, and naturally so.

Most social theory could explain Galton's results better than he did, but, by adding science to the biases of the day, he captured the public imaginary. His psychological studies of genius have been crucial to bell curve theories of individual differences that have turned schools into the early socialization and measurement arm of class and race structures in the United States (Henry, 1963; Bowles and Gintis, 1974; Gould, 1995). There were alternatives then, as now, but the bell curve mentality did not and does not go away. Galton's *Hereditary Genius* was published in 1869, the same year John Stuart Mill, in the *Subjection of Women*, ruled out biological constraints on the achievements of women, and Leo Tolstoy, in *War and Peace*, showed that the genius of military generals was best found in those who cowered in the right place at the right time before claiming victory (Latour, 1988). Alternatives available to Galton and his readers then, as now, were excellent, and for the last 40 years, every new toll of the bell curve of individual differences—with geniuses at one end and people of color at the other—has been devastated by empirical critique. What is left out at first is obvious: the environment in which organisms grow, the nurture that brings nature to bloom. What is less obviously left out is more crucial: It's us. It's Galton's readers, not his breeders, who are at fault. Genius is an attractive position at the top of a hierarchy of positions in which we must take our place. Genius sells, and we buy. Genius creates stupidity and, somewhere between, the rest are put into place. There is enough injustice in this intelligence game for many[3] to call for a new society that makes better use of the wisdom of its people.

Genius, Confrontations With

Addison identified the heroic, individual genius in the tensions of his times, and the same tensions soon spurred critiques. While the genius figure was still being hammered into shape, Henry Fielding (1707–1754) laughed at how much reality had to be ignored to make believe that genius is a constant, that a genius is always a genius, morally and intellectually so. While facing a new nation in search of an intellectual agenda, Ralph Waldo Emerson (1803–1882) formulated what a genius might be in a democratic society. With George Plekhanov (1857–1918), we are afforded less luxury and asked to put individual grandeur and genius back into the flow of the history that made achievement possible. These three points of view continue as alternatives to present arrangements.

Laughing at genius: To a growing list of who is a genius, Fielding added Jonathan Wild for contrast. Wild was a real person about whom Fielding wrote a fiction. The real Wild was a prominent criminal who had been hanged in 1725. The novel's Wild is no less a rogue, but an ingenious one:

> He was scarce settled at school before he gave marks of his lofty and aspiring temper; and was regarded by all his schoolfellows with that deference which men generally pay to those superior geniuses who will exact it of them. (1947, p. 9)

Fielding shows that being one step ahead of others should not be confused with wisdom or virtue:

> We must endeavor to remove some errors of opinion which mankind have, by the disingenuity of writers, contracted: for these, from their fear of contradicting the obsolete and absurd doctrines of a set of simple fellows, called in derision, sages or philosophers, have endeavored, as much as possible, to confound the two ideas of greatness and goodness: whereas no two things can possibly be more distinct from each other, *for greatness consists in bringing all manner of mischief on mankind, and goodness removing it from them.* (1947, p. 2; emphasis added)

Wild was supported by both the great mental agility required of a successful con man and a cast of supporters waiting for the spoils of his arts.

Fielding makes full fun of Wild's capacities:

> He was wonderfully ready at creating by means of those great arts which the vulgar call treachery, dissembling, promising, lying, falsehood, etc., but which are by great men summed up in the collective name of policy, or politics, or rather pollitrics; an art of which, as is the highest excellence of human nature, perhaps our great man was the most eminent master. (1947, p. 76)

Frieden reports that Fielding also allows, in *Tom Jones*, for genius as " 'a quick and sagacious Penetration into the true Essence of all the Objects of our Contemplation' " (1985, pp. 71–72). And why should not he? We could use more such moments—if only we could recognize them for sure, if only we were organized to make good use of them, if only we had an institutional program for their leading to the enhancement of all. Too bad Fielding did not offer a theory that confronts genius, or the position of genius, as an incomplete and sometimes dangerous idea. Others have since given it a try.

Democratizing genius: Can the word genius be used to help build a world in which genius might flourish? Let us call this Emerson's problem. He had read Smith, but his ideas on genius came from Plutarch, Goethe, and, half by contrast, from his long-term correspondent, Thomas Carlyle. Emerson's genius is a *fully attractive* character that celebrates less the individual bearer of genius and more the people who delegate a representative for the secret wonders of their minds. The term *fully* allows how much the genius takes from those around, not as in *fully stuffed* by their biases, but as in *fully representative* of their wisdom: "the greatest genius is the most indebted man" (Emerson, 1995; p. 127). *Attractive* allows how useful a genius might be to all around: the great poet gets praise for "appraising us not of his wealth but of the common wealth" (1990, p. 204). *Representative* is the appropriate political term for Emerson's genius. He was writing for a new democracy promising individuals conditions of growth that would in turn allow further growth for all (Dewey, 1903; J. McDermott, 1986; Shklar, 1990). Emerson's genius, as representative, seeks new connections, not to stand above others, but to go deeper into what joins them.[4] Self-reliance, however much a flag for the conformist individualism of today's consumer, is, for Emerson, an aversion to conformity:

> Society is a joint-stock company, in which the members agree, for the better securing of his bread to each share holder, to surrender the liberty and culture of the eater. The virtue in most request is conformity. Self-reliance is its aversion. (1990, p. 151)

Genius has an aversion to the artificial, the arbitrary, and the imitated. It must be tuned to the cultural surround, for "life lies behind us as the quarry from whence we get tiles and copestones for the masonry of to-day" (1990, p. 91). But genius must also move on and in, into the labor and dignity of life: "The mind now thinks, now acts, and each fit reproduces the other. When the artist has exhausted his materials, when the fancy no longer paints, when thoughts are no longer apprehended and books are a weariness—he has always resource *to live* . . . Thinking is a partial act." In the living, genius can "comprehendeth the particular natures of all men. Each philosopher, each bard, each actor has only done for me, as by a delegate, what one day I can do for myself . . . The man has never lived that can feed us ever. The human mind cannot be enshrined in a person who shall set a barrier on any one side to this unbounded, unboundable empire" (1990, p. 96).[5]

In *Representative Men*, Emerson takes from the usual pantheon (Plato, Shakespeare, Montaigne, Goethe), but surprises with two new genius entries: Swedenborg and Napoleon. Later, he complained about his choices, that he should have included "the common farmer and laborer":

> Many after thoughts as usual, . . . and my book seems to lose all value from their omission. Plainly one is the justice that should have been done to the unexpressed greatness of the common farmer and laborer. A hundred times have I felt the superiority of George and Edmund and Barrows, and yet I continue the parrot echoes of the names of literary notabilities and mediocrities. (Emerson, 1982, p. 406)

The entry on Napoleon—albeit biographically bumpy[6]—shows genius as representative and delegate, as a moment among minds. Take these two examples of Napoleon: first, the man of the people:

> though there is in particulars this identity between Napoleon and the mass of the people, his real strength lay in their conviction that he was their representative in his genius and aims, not only when he courted, but when he controlled. (Emerson, 1995, p. 162)

and then the man of democracy:

> I call Napoleon the agent or attorney of the middle class of modern society; of the throng . . . He was the agitator, the destroyer of prescription, the internal improver, the liberal, the radical, the inventor of means, the opener of doors and markets, the subverter of monopoly and abuse. (Emerson, 1995, p. 169)

For Emerson, Napoleon's genius fed "the native appetite for truth." Genius is close to the desires and knowledge of the people. "Bonaparte knew better than society; and moreover, knew that he knew better." Bonaparte knew better only what the people already knew: "I think all men know better than they do; know that the institutions we so volubly commend are go-carts and baubles; but they do not trust their presentments" (Emerson, 1995, p. 166). And again: "As to what we call the masses and the common men;—there are no common men" (1995, p. 22). Emerson's genius is of the people.

Revolutionizing genius: It is not easy to make a world safe for genius—or from genius. Within 50 years of Emerson's great "opener of doors and markets," a new Bonaparte, a nephew, also claiming the will of the people, ascended to a renewed throne of France. Descendents of the heroes of the French Revolution and the liberating armies of Emerson's Napoleon conspired to refurbish the throne their predecessors had destroyed, a situation ironic enough for Marx's famous diagnosis:

> Hegel remarks somewhere that facts and personages of great importance in world history occur, as it were twice. He forgot to add: the first time as tragedy, and the second as farce . . . The same caricature occurs in the circumstances attending to the second edition of the eighteenth Brumaire! Men make their own history, but they do not make just as they please; they do not make it out of circumstances chosen by themselves, but under circumstances directly encountered, given, and transmitted from the past. The tradition of all the dead generations weighs like a nightmare on the brain of the living. (Marx, 1963, p. 15)

Just when a genius can make a difference, people can close ranks. Just when monopoly and abuse can be subverted, people can create a counter force. The conditions, as they say, were not ripe. The usual definition of genius has him or her—nah, just him in the usual definition—enough ahead of his, uhhm, or her, time to ripen the conditions of participation. The genius should run ahead of society, but must come from that same society. How to think about a genius raised by the times, running ahead of the times, and reinserting advances into the evolution of the times? As phrased, this is the stuff of heroism. Only a great man can do it as phrased. As phrased is key. Plekhanov rephrases:

> A great man is great not because his personal qualities give individual features to great historical events, but because he possesses qualities which make him most capable of serving the great social needs of his time, needs which arose as a result of general and particular causes. (Plekhanov, 1969, p. 176)

For Plekhanov, the great person—the heroic genius—is a great beginner who reports inevitably to the most pressing local constraints and possibilities. Most beginnings end before their time; some go beyond expectations. Plekhanov had likely read Marx and Engles's account of the painter, Raphael: whether he "succeeds in developing his talent depends wholly on demand, which in turn depends on the division of labour and the conditions of human culture resulting from it." It is not as important to track the persons who made the first effort as to keep it moving, to capture the flow of history and to ride it powerfully to where it is inevitably going. Like other forms of labor, learning does not happen in a vacuum. Learning does not go to market alone (Lave and McDermott, 2002). Freedom requires doing as much as possible with/for inevitability:

> Everything depends upon whether my activities constitute an inevitable link in the chain of inevitable events. If they do, then I waver less and the more resolute are my actions . . . This is precisely the psychological mood that can be expressed in the celebrated words of Luther: "Here I stand, I can do no other," and thanks to which men display the most indomitable energy, perform the most astonishing feats. Hamlet never knew this mood; that is why he was only capable of moaning and reflecting. And that is why Hamlet would never have accepted a philosophy according to which freedom is merely necessity transformed into mind. (1969, p. 142)

This phrasing deserves play: Intelligence "is merely necessity transformed into mind." Revolutionary ideas are "merely necessity transformed into mind." Genius "is merely necessity transformed into mind." This is no laughing matter. Fielding would not laugh at a genius so tightly tied to reality. Working genius in an Emersonian democracy articulates "necessity transformed into mind." Plekhanov has a demanding sense of genius, one that stands, by necessity, against itself, against its own conditions. Genius has to make a difference in the flow of history. It has no time for rewards. The term genius cannot refer to what kids do on a test, nor academics in a book, nor scientists in a lab. Genius must grow from harsh realities it in turn transforms. Plekhanov would like Emerson's remark: "Genius is not a lazy angel contemplating itself and things. It is insatiable for expression. Thought must take the stupendous step of passing into realization" (Emerson, 1894, p. 40). Intellect is not the whole assignment. For Plekhanov, genius must change the world faster than the world can recover, before tragedy gets turned into farce by ignoring the inevitable.

Genius, Abuses Of

Confrontations with genius have not taken center stage with their enemy. Subtle calls for building a society that can take advantage of genius usually come in second to easier tasks like selling potential genius in the marketplace. A friend critiqued the idea of creativity by showing me a page in a phonebook listing companies under the word Creative. Her question: If it makes good business to sell blinds by calling the company "Creative Blinds," can creativity be an interesting analytic category? Her answer: Nah!. Emerson's answer: "We have a juvenile love of smartness, of showy speech. We like faculty that can rapidly be coined into money, and society seems to be in conspiracy to utilize every gift prematurely, and pull down genius to lucrative talent" (1894, p. 52). My answer: Better to treat received theories of creativity and genius as phonebook advertisements.

If the genius figure has been more about marketing than learning, what is the harm? It seems a gentle fiction: some get more celebrated than they should, others less than they deserve. The numbers are small and it puts learning—not gender, skin color, or wealth—at the door to institutional access. Why should we worry about genius? Answer: Measured calibrations of the mind are crucial to how gender, race, and class borders get maintained. How various groups align with genius shows an invidious relation to the distribution of rewards. Who gets left out or hurt by the category? The most obvious and best-researched example is the exclusion of people of African descent. The three cases that follow are comparatively minor, but perhaps in that way more revealing: women geniuses are suppressed, the Jews of Europe are celebrated and cursed for their genius, and American schoolchildren encounter genius as the measure of what they are not.

Gender and genius: Madam Curie is on everyone's genius list. Jane Austen has made it of late. I insist on Hildegaard of Bingen and Toni Morrison, Gertrude Stein perhaps, many of my friends, my daughters, and Yosano Akiko (early twentieth-century Japanese poet). Dinner conversations have brought names I would not have thought of: Anne Frank and Helen Keller, for example. And why not, but where is everyone else? The distribution is unfair, but how does it work?

Two critiques have emerged. One is that women have been strategically kept from participation in the arts and sciences, and inherent potential geniuses were suppressed. Another is that women have been producing works of genius, but men have refused, or did not know how, to recognize them. Evidence abounds for either case, for women have often had small public lives, and reduced access to genius games, because of institutional barriers. The invention of a special kind of genius (Alaya, 1977) for women only made things worse as men quickly claimed mastery of the new characteristics as well. Christine Battersby (1988, p. 113) offers Otto Weininger's early twentieth-century version of the feminine and intuitive male genius that had developed across the nineteenth century:

> The man of genius possess, like everything else, the complete female in himself; but woman herself is only a part of the Universe, and the part never can be the whole; femaleness can never include genius.

Women authors had to deal with this as they developed characters in their books—or in their own lives; Mary Wollstonecraft, Germaine de Stael, Mary Shelley, and Elizabeth Barrett had to fashion a vision with such conceptual tools. Genius is a competitive issue on which men have been combative. Whatever means developed for women to take their place, men have countered. Silly arguments have kept the genius inside the men's clubhouse. The following are borrowed from Battersby:

- "my coarse imagination has never been able to imagine a creative genius without genitals"—Johann Georg Hamann, 1760 (make that male genitals, says Battersby);
- "woman, in short, has an unconscious life, man a conscious life, and the genius the most conscious life"—Thomas Carlyle, 1840;
- "every genius born a woman is lost to humanity"—Stendhal;
- "with woman the powers of intuition, of rapid perception, and perhaps of imitation, are more strongly marked than in man; but some, at least, of these faculties are characteristic of the lower races"—Charles Darwin, 1871;
- "woman attains perfection in everything that is not a work: in letters, in memoirs, even in the most delicate handiwork, in short in everything that is not a métier"—Nietzsche, 1883.

To laugh at this foolishness cannot end the critique. To double the number of geniuses by including women, while an essential first step, cannot end the critique either. The critiques leave genius intact—suppressed or ignored in women, but intact. This is not the right choice. The kind of hyper-intellectual acrobatics that Weininger or Carlyle engaged in to negate the genius of women will not go away. In calling for female geniuses, we need more than numerical parity, more than more of the genius varieties males have been producing. The problem is that any confrontation with genius is a confrontation with male-defined genius. Battersby identifies five strands of meaning that have run through modern genius. The first four are loaded with male essentials, but the fifth is useful. The four are genius as personality, mode of consciousness, concentration of energy, and intelligence. The fifth is genius as a social activity. It confronts the first four and answers the question of how to think about genius as a moment more than a position, as an event more than a person:

> A female genius is not some kind of elite being, different from other (ordinary) women, nor one with a great "potential for eminence." A female "genius" is, instead, a woman who is judged to occupy a strategic position in the matrilineal and patrilineal patterns that make up culture. Her work must be seen to have a worth and importance that extends beyond mere popularity or influence. (Battersby, 1989, p. 157)

Women kept from genius games can be joined by cultural and racial groups accused of not being smart. Cultural anthropology is a long defense of the capacities of Others erased by dumb theories of intelligence. Battersby wants more women to be cited as geniuses, and she wants them to change the criteria of genius. A look at a minority people celebrated for genius shows she is onto something.

Race and genius: There are two ways to be hurt by theories of exceptional intelligence: one is to be thought lacking, the other to be caught having too much. The first insult has fallen on "primitive" societies in general, and people of African descent in particular. This has led to the cry of ethnographers pointing to intelligence where others—usually psychologists—had not seen it. The second insult has fallen on people of Jewish descent in Europe and, presently, Asian Americans. These are old story forms, retold with new tools. Confrontations with genius differ by theoretical trend and economic circumstance. The physical anthropology changes and so do the psychological measures of behavioral tendencies, but the results are consistent. If karma is a name for conditions always already there, anti-Semitism is karma to Europeans, and color racism is karma to Americans. A study of "smart Jews" shows how theories of genius fit a racial karma.

Having recently built a category for a genius and long before a negative position for Jews in Christianity, Europe was ripe to combine them.[7] High-profile Jewish success in the arts and sciences had to be explained. The first step was easy: Jews were smarter. This fit the eugenic logic of nineteenth-century evolutionary theory: Cleansed by centuries of persecution, Jews should be the most adaptive and intelligent people. With Jewish genius on a pedestal, what of the forces that made anti-Semitism? Cultural categories give, and they take away. Celebration turned into degradation reveals the original function of a category. Jews were lifted onto a pedestal by one reading, and taken off by a next. Gilman (1995) shows the European "yes, but" hand-wringing over Jewish genius: "Yes, they are" and "No, they're not." Or "they are a little, but not . . ." Consider some claims/counterclaims:

- genius, yes, but more in poetry and self-expression than in drama, which demands a reciprocity of perspective (Russian semiticist Daniel Abromovich Khvol'son, 1870s);
- genius, yes, but in practical activity, not in higher quality pursuits (Italian Jewish forensic psychiatrist Cesare Lombroso, 1894);
- genius, yes, but at the price of great nervousness and mental instability (French historian Anatole Leroy-Beaulieu, 1893);
- genius, yes, but at the price of avoiding physical labor (German historian Werner Sombart, turn of the century);
- genius, yes, but with more variability in the population; more geniuses perhaps, but lower average intelligence (Canadian psychologist Carl Brigham, 1923).

For every myth is a counter myth. For every distinction made, is another distinction unmaking it. The myth of genius makes individuals special, and, once special, new distinctions can counter the specialty. A list of people theorizing Jewish genius overlaps significantly with a list of people theorizing female genius. Genius distinctions were everywhere from 1860 to 1940, each one fodder for a next compulsion of the political order. Battersby tells a genius story with a positive outcome pulled from misery. Hannah Hoch was an artist in the Dadaist movement in the 1920s. Years later, the Nazis were destroying "degenerate" artists with the same relish they were arresting Jews. Hoch hid a stockpile of Dadaist materials around her home near Berlin. She was investigated by authorities and dismissed. Because she was neither a man nor a Jew, went the account, she could not have been enough of a genius to require extermination:

> Not an "Outsider"—only an "Other" that disrupts *Kultur* from *the margins within*—she seemed insignificant. Neither Jew nor genius, this female otherness allowed her to merge into the scenery . . . in much the same way as she later blended into the background in the histories of art. (1989, p. 144; emphasis in original)

Under the worst conditions, systematic inequalities in genius sighting can make things better.

Schoolchildren and genius: If genius has been a political issue in Europe since 1700, it did not became a political staging ground in the United States, until the twentieth century, when it was tied to theories of individual differences, IQ scores, school policy, and eugenics. The source was Galton, not Emerson, and the nexus was Louis Terman, who thought social progress could be quickened by using psychological tests to get the right persons for the right jobs. The influence of Termanal genius stretches to the present: Jews and Asians are statistically significant smart groups, and African Americans are a constant source of bell-curved reasons for whites to have the best jobs. The cultural fabrication of differential intelligence requires always two ends; both smart and dumb must be visible. The violence invited by the American theory of genius is not yet tallied. Geniuses are not trapped in ghettoes, and their counterparts at the bottom of the bell curve are so cleverly kept from mobility that problems are easily ignored, silenced, and muted (Pollock, 2004). The attribution of genius makes possible the attribution of stupidity. The genius figure helps to hold the system together.

The image of a lockstep stairway to cognitive excellence is strong in American blaming practices. If a child seems unable, school personnel can be heard saying, "The child isn't smart enough. Everyone has limits." Parents report the other side of the coin, saying, "My child is smart, but doesn't get help in school." Smart is the acknowledged key. All roads return to it. Life trajectories are explained by it, and there is a label for everyone. From gifted and talented to LD and retarded, the system rarely gets questioned. The person who objects is met with Einstein as if to challenge the use of "smartness" in grade schools is to deny individual differences. How could anyone deny individual differences? Not denying that everyone is different enough to be interesting does not mean we know how to identify individual differences fairly, or which ones count the most, or how to build institutions for making the most of them. *Individual differences on arbitrary tasks often unconnected to what has to happen in life should not be the measure of any new generation.*

As one end of an ill-conceived and badly applied theory of individual differences, genius silently aids inequality in American schools by delivering success and failure correlated with race and class. No one has to be a genius, or even to be called a genius, for the system to work. A nice idea, genius—Diderot thought so—but it has not kept good friends. It seems grounds for good things: individual achievement, room for new ideas, rewards for those most able to help. A good idea, but it has also encouraged arrogance, sexism, racism, class hierarchy, crass individualism, and static ideas of inherited intelligence. We need new ways to engage genius, to rediscover the promise of vision and creativity across a community. Pascal, Emerson, and Plekhanov have shown the dangers of locating genius in the lone person. They ask instead that we build societies that would allow a kind of genius we might want to get behind, the kind of genius who could represent us.

Summary

An inquiry into genius is not about how one person gets better ideas than others, nor about how one person jockeys for position better than others (although biographies attest that most acclaimed geniuses have pulled every social structural trick imaginable to help others to the illusion[8]—maybe even Pascal before he sought humility). These two questions are initial to a wider account of how genius positions are concertedly made possible and formed in a historical setting. Analytically and politically, genius is an opportunity in a set of opportunities set in a daily round of celebration, degradation, and interpolation, all performed by real people in systems of well-structured and irremediably emergent positions. It is not enough to describe the mental world of any genius, nor enough to show that the mental world of genius is quite ordinary. It is more exciting to show how every kind of person—every kind of sensuous, engaged, hoping, wishing, thinking, and conniving person—is also a calibration in a moving system of positions made up of materials gathered from here and there in the service of forces often unseen and unnamed (Holland and Lave, 2002). Any inquiry into living and learning in America must offer an account of the organization of desires, languages, budget lines, and institutional necessities that together arrange for so many to get tagged as illiterate, disabled, slow, inarticulate, at risk, failing, and all that as a contrast case for the smart, gifted, successful and ingenious (Mehan, 1996; Varenne and McDermott, 1998; Rapp and Ginsberg, 2001).

Kinds of person described by presumed mental states are the calibration of moment. Poised between a rhetoric of democracy and the demands of a competitive capitalism, the genius, the mentally ill, the dumb, the autistic, and the hyperactive are in the service of a hierarchic political order. Sub rosa, they sort people by gender, race, and class. Whatever else they might be, *named mental kinds of person, genius and all its counterparts, are irrevocably calibrations in the ups and downs of advantage and the ins and outs of access.*[9] People were not always forced to look smart as the only way to gain access to society. Looking smart is offensive in most cultures (Mitchell,

1988; Hori, 1994; Basso, 1996). Intelligence, if noticed at all, should be more used than displayed. Gradations of intelligence do not have to serve the competitive demands of the market, but degradations of intelligence are made for the job. Crude measures of the mind have been given purchase power. Genius has become too expensive.

Acknowledgments

Grey Gundaker is the friend who taught me to read the NYC phonebook. Vicky Webber showed me Battersby (1988) and Elisabeth Hansot pointed to Valéry (1973). Shelley Goldman, Dottie Holland, Nate Klemp, and Jean Lave offered a critical reading, and audiences at Stanford and Teachers College made suggestions. An abbreviated version of the first half of this chapter appeared in McDermott (2004a).

Notes

1. Addison is often cited as the first analytic take on genius. There is also Juan Huarte from 1594 (1946), but his work does not move forward to eighteenth-century discussions.
2. Smith on Pascal: a "whining and melancholy moralist . . . perpetually reproaching us with our happiness, while so many of our brethren are in misery, who regard as impious the natural joy of prosperity . . ." (1976, p. 139).
3. In 1808, Henri Grégoire (1996) issued a manifesto we should not ignore: "Irishmen, Jews, and Negroes, your virtues and talents are your own; your vices are the work of nations who call themselves Christian. The more you are maligned, the more these nations are indicted for their guilt" (1996, p. 39). Replace the words "Irishmen, Jews, and Negroes" with any pariah group, and replace "nations who call themselves Christian" with the self-assessed smart. There it is: The more those left out are maligned as stupid, the more the intelligent should be indicted for their part—Galton be damned.
4. Emerson directly influenced Nietzsche's genius. Compare his gentle, "Genius is always sufficiently the enemy of genius by over-influence" (1990, p. 87), with Nietzsche's edgy, "A people is the detour made by nature to arrive at six or seven great men. Yes, and then to get around them" (1997, p. 66)—the same point, to a variant purpose. Past genius can be today's burden, but Nietzsche's genius lords over people until pushed aside. Emerson's is more integral and worth recovering.
5. Gertrude Hughes (1984, p. 82) captures the reflex between bounded and boundless in Emerson's genius: "The sublimely apt yet utterly homely ratio whereby landscape is to sight as the psyche's terrain is to

genius' insight works this way, for it makes us re-evaluate the ordinary fact that landscape surrounds us. The simile not only compares the achievements of genius to something we experience every day, but also reminds us that we do not realize we experience it every day. Thus Emerson is not just using the known to make the unknown accessible. In the same stroke by which he does that he also manages to 'make the visible a little hard/To see' as Wallace Stevens was later to write that the worthy poet must do."

6. Emerson put quotation marks around what Napoleon had said—and around what he might have said, could have said, or, to Emerson, should have said. Schirmeister (1993, p. 223) says Napoleon likely never said: " 'My son cannot replace me. I could not replace myself. I am the creature of circumstances' " (Emerson, 1995, p. 155). The less accurate Emerson is on Napoleon, the more he reveals of himself.

7. This summary of Gilman (1995) is adapted from McDermott (2004b).

8. Paul Valéry noted a tie between self-celebration and genius: "What they call a superior man is a man who has deceived himself. To be astonished at him, one must see him—and to be seen, he must show himself . . . In exchange for the public's dime, he gives the time required to make himself noticeable, the energy spent in conveying himself, preparing to satisfy someone else. He goes so far as to compare the crude sport of fame with the joy of feeling unique" (1973, p. 9).

9. In 1892, about midway between Adam Smith's warning and the present fetish for displays of intelligence, Charles Sanders Peirce summarized the coming use of intelligence in a twentieth-century democracy under capitalism:

> Political economy has its formula of redemption, too. It is this: Intelligence in the service of greed ensures the justest prices, the fairest contracts, the most enlightened conduct of all the dealings between men, and leads to the *summum bonum*, food in plenty and perfect comfort. Food for whom? Why for the greedy master of intelligence. (1996, p. 271)

In 1868, Peirce (1955) critiqued easy psychologizing, intelligence being a latter-day token of type. In 1938, John Dewey, heir to Emerson and Peirce, complained that twentieth-century theories of human nature are psychological in ways that support the "intrinsic and necessary connection between democracy and capitalism . . . for it is only because of a belief in a certain theory of human nature [of the self-involved, strategic consumer] that the two are said to be Siamese twins, so that the attack upon one is a threat directed at the life of the other" (2001, p. 160).

References

Addison, Joseph (1711). [no title] *Spectator* 160 (September 3, 1711).
Alaya, Flavia (1977). Victorian Science and the "Genius" of Woman. *Journal of the History of Ideas*, **38**, pp. 261–280.
Balzac, Honoré de (2002/1832). *Louis Lambert.* McLean: IndyPublish.com.
Basso, Keith (1996). *Wisdom Sits in Place.* Albuquerque: University of New Mexico Press.
Battersby, Christine (1988). *Gender and Genius.* Bloomington, IN: Indiana University Press.
Bourdieu, Pierre (2000). *Pascalian Meditations.* Stanford, CA: Stanford University Press.
Bowles, Samuel and Herbert Gintis (1974). *Schooling in Capitalist America.* New York: Basic Books.
Cailliet, Emile (1961). *Pascal: The Emergence of Genius.* 2nd edn. New York: Greenwood Press.
Dewey, John (1903). Emerson, the Philosopher of Democracy. *International Journal of Ethics*, **13**, pp. 405–413.
Dewey, John (2001/1938). Democracy and Human Nature. In Stephen J. Goodlad (ed.), *The Last Best Hope: A Democracy Reader*, pp. 159–175. San Francisco: Jossey-Bass.
Diderot, Denis (1964). *Rameau's Nephew and Other Works.* Indianapolis: Bobbs-Merrill.
Dieckmann, Herbert (1941). Diderot's Conception of Genius. *Journal of the History of Ideas*, **2**, pp. 151–182.
Emerson, Ralph Waldo (1894). *Natural History of Intellect and Other Papers.* Boston: Houghton, Mifflin, and Company.
Emerson, Ralph Waldo (1982/November–December 1849). In Joel Porte (ed.), *Emerson in His Journals*, pp. 405–406. Cambridge: Harvard University Press.
Emerson, Ralph Waldo (1990). In Robert Richardson (ed.), *Selected Essays, Lectures, and Poems.* New York: Bantam.
Emerson, Ralph Waldo (1995/1850). *Representative Men.* New York: Marsilio Publishers.
Engell, James (1981). *The Creative Imagination.* Cambridge: Harvard University Press.
Fielding, Henry (1947/1743). *The History of the Life of the Late Mr. Jonathan Wild, the Great.* London: Hamish Hamilton.
Frieden, Ken (1985). *Genius and Monologue.* Ithaca, NY: Cornell University Press.
Galton, Francis (1872). Statistical Inquiries into the Efficacy of Prayer. *Fortnightly Review*, **12**, pp. 125–135.
Galton, Francis (1962/1892). *Hereditary Genius*, 2nd edn. New York: Peter Smith.
Gilman, S. (1996). Smart Jews in Fin-de-Siecle Vienna: Hybrids and the Anxiety about Jewish Superior Intelligence—Hofmannsthal and Wittgenstein. *Modernism and Identity*, **3**, 2, April, pp. 45–58.

Goldmann, Lucien (1964/1952). *The Hidden God*. New York: Routledge and Kegan Paul.
Gould, Stephan J. (1995). *The Mismeasure of Man*. 2nd edn. New York: Norton.
Grégoire, Henri (1996/1808). In Thomas Cassirer and Jean-Francois Briere (eds.), *The Cultural Achievements of Negroes*. Amherst, MA: University of Massachusetts Press.
Henry, Jules (1963). *Culture against Man*. New York: Vintage.
Holland, Dorothy and Jean Lave (2002). History in Person: Introduction. In Dorothy Holland and Jean Lave (eds.), *History in Person*, pp. 3–32. Cambridge: Cambridge University Press.
Hori, Victor (1994). Learning in a Rinzai Zen Monastery. *Journal of Japanese Studies*, **20**, pp. 5–36.
Huarte, Juan de San Juan (1946/1594). *Examen de Ingenios para las Ciencias*. Buenos Aires: Espasa-Calpe Argentina.
Hughes, Gertrude (1984). *Emerson's Demanding Optimism*. Baton Rouge, LA: Louisiana State University Press.
Jaffe, Kineret (1980). The Concept of Genius: Its Changing Role in Eighteenth-Century French Aesthetics. *Journal of the History of Ideas*, **41**, pp. 579–599.
Koyré, Alexander (1952). An Unpublished Letter of Robert Hooke to Isaac Newton. *Isis*, **43**, pp. 312–337.
Kroeber, A. L. (1944). *Configurations of Culture Growth*. Berkeley, CA: University of California Press.
Latour, Bruno (1988). *The Pasteurization of France*. Cambridge, MA: Harvard University Press.
Lave, Jean and Ray McDermott (2002). Estranged Learning. *Outlines*, **4**, pp. 19–48.
Marx, Karl (1963/1852). *The Eighteenth Brumaire of Louis Bonaparte*. New York: International.
McDermott, John (1986). *Streams of Experience*. Amherst, MA: University of Massachusetts Press.
McDermott, Ray (1988). Inarticulateness. In Deborah Tannen (ed.), *Linguistics in Context*, pp. 37–68. Norwood, NJ: Ablex.
McDermott, Ray (2004a). Materials for a Confrontation with Genius as a Personal Identity. *Ethos*, **32**, pp. xx–xx.
McDermott, Ray (2006). *Situating Genius*, in press.
Mehan, Hugh (1996).The Construction of an LD Student. In Michael Silverstein and Greg Urban (eds.), *Natural History of Discourse*, pp. 253–276. Chicago: University of Chicago Press.
Mitchell, Timothy (1988). *Colonizing Egypt*. Cambridge: Cambridge University Press.
Murray, Penelope (ed.) (1989). *Genius: The History of an Idea*. London: Blackwells.
Nahm, Milton (1965/1956). *Genius and Creativity: An Essay in the History of Ideas*. New York: Harper.

Nietzsche, Friedrich (1997/1884). *Beyond Good and Evil*. London: Oxford University Press.
Oakley, A. (1994). *Classical Economic Man: Human Agency and the Methodology in the Political Economy of Adam Smith and J.S. Mill*. London: Edward Elgar Publishing.
Pascal, Blaise (1960/1670). *Pensées*. London: J.M. Dent & Sons.
Pascal, Blaise (1989a/1652). Letter to Queen Christina. In Richard H. Popkin (ed.), *Pascal: Selections*, pp. 30–32. New York: Macmillan.
Pascal, Blaise (1989b/1660). Three Treatises on the Condition of the Great. In Richard H. Popkin (ed.), *Pascal: Selections*, pp. 74–78. New York: Macmillan.
Peirce, Charles Sanders (1955/1868). Four Incapacities and Their Consequences. In Justus Buchler (ed.), *Philosophical Writings of Peirce*, pp. 228–250. New York: Dover.
Peirce, Charles Sanders (1996/1893). Evolutionary Love. In Morris Cohen (ed.), *Charles Sanders Peirce, Chance, Love, and Logic*, pp. 267–300. Lincoln, NE: University of Nebraska Press.
Plekhanov, George (ed.), (1969/1898). The Role of the Individual in History. *Fundamental Problems in Marxism*, pp. 139–177. New York: International Publishers.
Pollock, Mica (2004). *Colormute: Race Talk Dilemmas in an American School*. Princeton, NJ: Princeton University Press.
Rapp, Renya and Faye Ginsberg (2001). Enabling Disability: Rewriting Kinship, Reimagining Citizenship. *Public Culture*, **13**, pp. 533–556.
Rogers, Ben (1999). *Pascal*. London: Routledge.
Rothschild, Emma (2001). *Economic Sentiments: Adam Smith, Condorcet, and the Enlightenment*. Cambridge: Harvard University Press.
Schirmeister, Pamela (ed.) (1993). Introduction. In *Representative Men*. New York: Marsilio Publishers.
Screech, M. A. (1981). *Montaigne and Melancholy*. London: Penguin.
Shaw, Bernard (1928). *The Intelligent Woman's Guide to Socialism*. New York: Brentano's.
Shklar, Judith (1990). Emerson and the Inhibitions of Democracy. *Political Theory*, **18**, pp. 601–614.
Smith, Adam (1976/1759). *The Theory of Moral Sentiments*. London: Oxford University Press.
Smith, Adam (1988/1776). *An Inquiry into the Nature and Causes of the Wealth of Nations*. New York: Modern Library.
Valéry, Paul (1973/1896–1925). *Monsieur Teste*. Princeton, NJ: Princeton University Press.
Varenne, Hervé, and Ray McDermott (1998). *Successful Failure: The Schools America Builds*. Boulder, CO: Westview Press.
White, Leslie (1949). *The Science of Culture*. New York: Free Press.

Chapter 11

What You Don't Know about Diversity

Elizabeth Quintero

Diversity. What has it got to do with learning in twenty-first century schools? What does it have to do with all the pressures of standards, accountability, and budgets? Everything.

What you may not know is that diversity in education is much more than court cases on affirmative action and much more than multicultural literature and history. And those too. It's more than the fact that since the 1990s, many more communities in the United States have become home to people from all over the planet. And it is that too.

Diversity is personal. Diversity is communal. Schools and curricula are a composite of personal learning and teaching that occurs among students and their teachers. So what? I invite the reader to think about this "so what?" as the chapter continues. There, of course, is no easy or simple answer to the "so what?" question. The only nonanswer is not to think about it and not to do anything.

The Personal

In 2003, a teacher education student in a graduate program in New York City wrote about her personal memories of diversity as a young child:

> I came from India when I was four and started kindergarten when I was five. Neither one of my parents spoke English, so when I started school it was a difficult experience for me. I felt so different from the other children and wanted to quickly learn English so that I can be a part of their world. I would run home after school so that I could finally be in a world that was familiar to me, with language and customs I was an expert of. However, at the same time I was angry with my parents for not knowing and therefore not being able to teach me English. I felt very alone in my experience. When my mother would pick me up after

school I would beg her to not speak to me in Hindi. I was embarrassed about who I was.

What a shame for a child to be embarrassed about who they are because of the fear of rejection and because of the pressure to assimilate and build a new, acceptable identity for the new world they are a part of.

I am glad that I quickly realized the gifts of being different. And it is sad to say that it wasn't the teachers that helped me, nor my parents. It was the other kids at school. You will meet kids who will reject you right away because you're different and then there are those kids who are intrigued and want to learn from you and about you. That makes you feel special.

Another teacher education student in New York City, after going through a series of activities relating to family history and the influence of culture and language on literacy, wrote about her personal diversity perspectives:

I'm Chinese-American and in the Chinese community, I would be described as a jook-sing, in other words, a child born in America that has adopted many American customs and characteristics. Being Chinese was not something I was proud of when I was younger and because of that I failed to learn as much of the language as I could have, which I feel is a big part of being Chinese. Through this language I am able to have conversations with people who can fill me in on what being Chinese is all about. But because my ability is limited I am often shy with my relatives who only speak Chinese. I don't want to be branded a jook-sing although I feel I am one. I am proud to be Chinese but when people ask me to tell them details about Chinese culture I'm afraid that I may fall short in explaining the complexity of it all. This aspect of me is important; it connects me to my family and their values. I am the way I am because of my ethnicity. I have accepted some of these beliefs and values as my own but I have also rebelled against some values that clash with my understanding of the world. I feel the latter aspect of myself can be attributed to my Americanized side.

I don't tie my ethnicity to a particular place. I bring it wherever I go. However, I will say this, I feel my Chinese-ness the most if I am the only Chinese person in a particular place. I rarely feel this way in New York because of its diversity. Mostly I carry both plates in my hands wherever I go but the people and the place will determine which one will be heavier.

Another student in the same class wrote about the complexity of her family:

When I think of my ethnic identity, honestly I think of so many things I don't know where to start. My family members come from all over the place actually, so I don't really know which ethnic group I identify myself with most. My mother is American by birth, as is her mother. But, my grandmother grew up in Mexico, as did my grandfather. My grandmother's family was from Mexico. My grandfather's father was from Greece, his mother from Mexico. My father, born in Mexico, is of European descent. His father was born in Stockholm, his mother in San Antonio, right next to the Alamo. My grandmother also has Swedish and German ancestry. And while he is not Latino, my father grew up in Venezuela. These are all of the things I think of when I think of my ethnic background. I tend to relate more to the Hispanic side of my family, because I grew up in San Antonio where there is such a large Hispanic population. I guess that would be the place my ethnic identity is tied to. Everyone in my family speaks Spanish; all my grandparents, aunts, uncles, etc. While the history of Mexican people has not played a large role in my upbringing, I know some of it, mostly what I learned in school and from my surroundings. People in San Antonio are very aware of the Mexican culture and this is shown in various ways throughout the city. Even though I am American, I relate to the Mexican culture most out of all of my backgrounds because it plays such a vital role in my everyday life. There is not a large Swedish or Greek community in San Antonio. The earliest things I remember about my cultural background and ethnicity are the traditions of my family. I did not know much about Swedish people, but I knew what they ate at Christmas time. I knew some of their manners from stories told to me by my father's parents. I knew about the Greeks and what kinds of things they did at Easter. My mother's parents would take us to the Greek Festival held each year at St. Sophia's Greek Orthodox Church downtown. This is how I was introduced to my ethnic background.

When I think of race and ethnicity, I realize how complex of an issue it is. Because of my background, I look white to most people who meet me. Many people are surprised to learn that I am Hispanic. While many Hispanics that I know have encountered racism before, I have not because I do not appear to be Hispanic. I have a hard time saying that I am either white or Hispanic. Both ethnicities are part of me, so I cannot say which I identify with more.

I hope this all makes sense to you. It seems like it all fits in my head. When I have thought about it before, I never thought I was confused about this. It seems like I am a bit, though. I think part of it is what I have learned about race and ethnicity and culture over the past few years. They are not things that are so easily defined, I guess.

Yes, diversity is complex for almost all of us. Carl Grant, at a small early childhood conference in Madison in 1991, asked me if I would tell my story for a book he was preparing (*Educating for Diversity*, a book I eventually used in some classes). He needed a Cuban American perspective of struggle and success. "I can't," I said. "Why not?" He was surprised. Astute observer, he was probably thinking of our 10-min conversation there in front of the fireplace after Lourdes introduced us. *Florida, Texas, Mexico, the surname. The names of your boys. Your underlying family focus in your work.* "No, I can't," I apologized. "I wasn't raised Cuban American. I wasn't raised anything." *Other than one of two sisters raised by a single mother, dark skinned with green eyes and curly hair, who was often silenced in her life.*

I told Carl Grant, "I'm a mongrel who is clearly different from the mainstream, but with only gotas (drops) of Latina, maybe Muscogee Creek, maybe black. I don't know. My family is one characterized by silences and not much historical memory. I can really not claim any identity other than me." He looked at me. "You could pass for several," he said with a twinkle in his eye. "There are more and more mixed ones like you coming up these days."

Diversity, the Personal Becoming Communal

So in what ways does this acknowledgment that diversity is personal relate to the community of the classroom, the community of a school, a neighborhood, and ultimately, the world community? The simple answer is that because diversity is personal, it is right there, up front and on the table, in most (if not all) human relationships. And frankly, with the exception of a hermit meditating on a hillside, a medical researcher in quarantine, and the rare human who has determined that he has no need for any communion with others, relationships are relevant to all learners, all humans.

What can I do about diversity in the communal sense? My issues of conviction and work revolve around paying attention to the strengths of parents and children. I believe educators and leaders in all arenas can learn more about how culture, language, and varying concepts of family affect child development, community development. I believe that this will ultimately improve our ability to live with each other with respect and peace. By briefly going back in time, I will highlight a few instances relating to why these issues became so important to me.

I grew up in Florida in the 1950s when it was okay to speak Spanish only at dance classes (in Tampa) in the weekends. When I was 16, I worked for Project Head Start in the summer of 1966 as an assistant teacher. We visited rural Black families in central Florida. After all these years, one of my most vivid memories is of the homes where the family members of various ages gathered on their front porches telling stories and "acting them out." I did not know then that one day I would learn that that was a classic example of both art and literacy. Later, I decided early childhood education was what I would do.

Yet, I had had friends in teacher education programs who told me about classes of boring texts and superficial memorizing of various "methods" of teaching and "recipes" for discipline. I guess, as a result of the combination of life experiences and personal philosophy, I wanted to study early childhood education from an alternative perspective. I went to visit the British Infant Schools and the Summerhill School in England. Fairly soon after arriving in London, I learned about a training program for preschool teachers that involved much practical experience in various inner-city neighborhood preschools. I plunged in and to my delight I was placed in a school in a Middle Eastern immigrant neighborhood.

Later, back in north Florida, I found work in a small preschool for 3- and 4-year-old children. After I finished my masters in education in early childhood studies, I extended my work to kindergarten teaching and literacy and science teaching at the elementary level. It was during this chapter of my life that I became very involved in asking the question, "What's left out . . . of my personal experience, of my education, and ultimately, my teaching?" This was the beginning of the conviction about which I talk now when I tell students to always ask questions about whose stories and opinions are left out of every textbook, every research study, and every news report. Diversity is often shadowed between the lines.

I was still learning about culture, language, teaching, and learning during the next decade when I lived in Mexico. After a few years, I went to New Mexico to work on a doctorate in early childhood/bilingual education. This was when Yetta Goodman and others—during the beginnings of the whole language movement—spoke of young children's play and communication as exemplifying the "roots of literacy." I realized that I had been studying and observing firsthand these roots of literacy in the children I had been working with for years, wherever they were, monolingual English-speaking children, African American children both in rural schools and in inner city schools, monolingual Spanish-speaking children in Mexico, Spanish–English bilingual children in Texas and New Mexico, or the Middle Eastern children in London. I had seen the "roots of literacy" as an integral part of what children do as they understand and take part in their world. Ironically, when families do not speak English, this "world" of language and literacy is often considered less than adequate and even deviant because it is different.

As I worked with children and parents in a variety of programs, I saw that every parent I met—from a diversity of circumstances, from difficult to comfortable—cared deeply about his or her child who was being entrusted to my care. I was developing the perspectives and commitments that would lead me in the directions that guide my work today.

All societies, as reflected in their schools, are a collage of strengths and barriers, voices seldom heard, and voices more often heard. When schools bring in the strengths of the families and their communities, in all their pluralistic complexities, the educational experiences are more effective. The strengths of the families and students may be valued and used by the schools or they can be ignored and wasted.

Schools can reflect the strengths of families. Unfortunately, they can also mirror reflection of public perceptions of the poor, thus perpetuating the tragic misconception of minority and poor children. Yet, schools can also be a place for change. As teachers learn to read children's texts more critically, they will help children learn to be discerning readers and writers of their own and other texts and explore through language new metaphors and structures that are challenging.

Now I know it is evident that I am a teacher and a qualitative researcher with a perspective about learning that frames the way I work and conduct research with all ages of learners from many different backgrounds. I research issues of education, families "at promise," language and critical literacy in the context of home culture and culture of learners' new learning environments in both schools and communities. A student I am working with, who is studying to become a teacher, wrote about ways her history and ongoing personal experiences with diversity had the potential to provide some meaningful bridges for students and families in the community of a school. She wrote at the beginning of her student teaching experience in Chinatown:

> On the first day of class, most of the parents and all the students in the entire elementary school gathered in the cafeteria on the first floor. As I waited for our class to arrive, a father approached me and began speaking in a Chinese dialect I regretfully could not understand. I explained that I could not understand and he apologized. Thinking back, I should've asked him in Mandarin if he spoke Mandarin because it is a fairly universal dialect. We could have made that important instant connection instead of creating distance.
>
> On another day, I walked a student down alone because he was the last one to copy down his homework. As we approached our meeting ground on the playground, the student's father asked me if I spoke Mandarin. I smiled and replied, "Yes!" The father continued on to express his concern about his son and homework. Basically, he never saw his son do homework; his son just played video games all the time. Last year, he found his son's desk drawer stuffed with papers and thought he might be doing the same this year. To sum up our conversation, I informed the father that we did indeed assign homework every day and I would make sure to check to see he had completed it each and every day. We discussed other topics and concerns but his main worry was his

son's homework. As they departed, the father asked me what my surname was (I'm not sure how to translate, but when you ask this question, it implies that you have a Chinese surname). I told him that I was mixed blood, but my mom's name is Lin. He then asked me if he should call me Lin Xiao Jie (Miss Lin) and I said, "of course." It felt really amazing to feel his trust in me and to have established this connection. So when my cooperating teacher invited me to the parent–teacher conference, I excitedly looked forward to meeting the other parents.

At the early morning meeting, I helped one of the parents translate my cooperating teacher's concerns, thoughts and answers to questions. By the end of the conference, . . . I was able to calm some of their fears and assure them their son or daughter was doing great and would excel. One mother was worried about her daughter needing speech therapy, so we visited the speech therapist together and I helped her translate that her daughter improved so much last year, therapy wasn't necessary this year. Needless to say she was very pleased. It made a big difference being able to speak their native tongue and I'm really glad I was placed in a bilingual school.

Another student, who is very knowledgeable about diversity issues, acknowledged that we always have more to learn. She reported to the class:

I had a chance to speak with my mother's hairdresser this weekend. Her name is N. and she is from Senegal. N. lived in Senegal most of her life. She has been in the United States for only two years. I was actually fascinated with her story, because she accomplished so much in just two years. Learning a different language as an adult is difficult and N. speaks Sengali and English. She works as a hairdresser and she is a student at the local community college. She says she is taking math and English courses. She said she's good in math but she's struggling with English. She said, "I get all A's in math, but English is too hard." N. works in Flatbush; many people from Africa work and live in this area. N. said she used to be embarrassed by her accent when she's speaking English. Now she says she is more confident. I feel confidence is an important factor when acquiring different languages. Schools and teachers have to support and encourage ESL students just like English speaking students. ESL students should be a part of the classroom community and not isolated from it.

Diversity in the Classroom

A friend and I studied effective literacy teachers (Rummel and Quintero, 1997) and found that teachers bring their past experiences, present values, and priorities into the schools. Their beliefs and life experiences cannot be separated from what they do in the classroom. Outstanding educators show an interest in and acceptance of many students' families, cultures, and differences.

We (Rummel and Quintero, 1997) found that teachers who support children and their cultural, linguistic context in school have some common approaches to pedagogy. They all exhibit a belief that it is their responsibility to find ways of engaging all their students in learning activity. They accept responsibility for making the classroom an interesting, engaging place. They persist in trying to meet the individual needs of the children in their classes, searching for what works best for each student. Their basic stance is a continual search for better ways of doing things.

A teacher in Brooklyn, New York, in a class studying curriculum, wrote in her response journal:

> Our cultural history is tied to literature. All aspects of history are incorporated into stories. These stories have been told over time and written down. Stories explore a multitude of perspectives about life and truth. Through these perspectives we gain a greater understanding of being part of a culture. As educators, we must present a wide variety of books to stimulate thinking. We must use books from each genre: poetry, traditional, fantasy, realistic fiction, informational, picture books, humor, predictable and multicultural. Teachers must follow the child's interests and present a collection of books that are intriguing.

Another student teacher suggested:

> Using literature and music are other ways that can expose children to educational material. Instead of reading the Disney's *Cinderella*, why not pick up a copy of the Korean version? Not only does it create a chance to expose children to different types of literature, but it can also lead to other topics like different foods, celebrations, music, clothing, and so on. In order to create well-rounded, critical thinkers we need to expose children to more than just chalk on a blackboard.

Yet another student teacher commented, "Within our use of language is our value system . . . if English is the only language accepted in school, this implies other languages are not as valued."

Finally, a group of teacher educations students were read the storybook *Madlenka* by Peter Sis. The story is about a little girl named Madlenka whose tooth is loose. She visits her friends and neighbors around her block to let them know. The book introduces each of them and shows a little about their culture, language, and customs. The illustrations display different symbols, monuments, and geographic locations that are representative of the countries mentioned in the book. In the beginning of the book there is a map of Manhattan that shows exactly where Madlenka lives, and at the end of the book is a map of the world, showing the locations of countries of the people that she interacts with without even leaving her city. After the reading, the assignment was to go out into a community near their current teaching placement and bring back information that could stimulate a tangible learning experience for various ages of students.

One student wrote after the experience:

> While on my Madlenka journey, it was almost impossible to escape visual art. I collected a huge collection of menus from all the restaurants in this neighborhood. I was attracted to all the drawings and small sketches on the covers. It isn't the type of art that will make it into a museum, but it does represent a small piece of the community. Graffiti seems to be the most abundant form of art in this neighborhood, and although it can be a nuisance at times, some of it is lovely and shows that someone took a lot of time and energy to put it there.

A student in another neighborhood in New York City wrote that she went on a walk around the block and wrote, "How amazing is it that all the ideas are right before our eyes and we don't see them unless we are asked a specific question exploration?" She then listed the ideas she got on her walk that would be useful in the classroom:

1. Go on a class scavenger hunt for different languages printed throughout your neighborhood.
2. Go on a class scavenger hunt for colors of the rainbows. Take photos and enlarge to create a "Class Rainbow" for the wall.
3. As part of a shape unit, dispense disposable cameras to groups of 2–3 children. Go on a shape hunt through the neighborhood. Develop and assemble shape books.
4. Go on a "sounds of the city" scavenger hunt. Use a recorder to amass a montage of the sounds of our neighborhood.
5. Invite parents and family members to come in to school with an artifact of cloth from their native culture. After spending some time discussing values of colors, hues, textures, and shapes use a variety of mediums to assemble a mosaic representing the many cultures in one small room. The result—"our Class Quilt"

Then she added:

> While I was walking, I stopped for a cup of coffee. I sat down at the counter and immediately took note of two young Asian children who were ripping up construction paper. I waited a few minutes and watched them with interest. Then, I asked them what they were doing. They told me that it was their great-grandmother's birthday and

"because she's old" she likes these kinds of old things. I asked them what they meant and they explained that you couldn't always go in to a store and buy cards or even writing paper, and the people in Tokyo invented a way to make paper, and that every year that is what they make their great-grandmother as a gift. One of the children pointed above the counter and showed me another card that they had made. Right there before my very eyes was a culturally rich history/geography/art lesson with a practical product!

Collective Learning about Diversity from the Real Experts—Families and Children

At a teacher development seminar for Head Start teachers with multicultural, multilingual programs in Minnesota, issues about making the early learning curriculum responsive to children's real lives, their strengths, and needs were addressed. There was much discussion and many suggestions were made in the groups about activities supporting Hmong, Lao, and Vietnamese children because these refugee groups had been in Minnesota for a number of years. When families developed trust in the programs, many of the parents had learned enough English to be able to assist the staff in curriculum development. Furthermore, storybooks and historical accounts of these groups had recently become available to educators.

Then, the refugees began arriving from Somalia. The teachers and staff talked about the number of vibrant, curious children and their respectful and quiet parents. The refugees came from a terrible war-torn reality and few knew just more than two or three words of English. One Head Start teacher stood up and said, "I have a story about what I learned from the Somali children about curriculum."

She explained that in her class of twenty 3- and 4 year olds that year, twelve of them were from Somalia. She related her attempts to gain the trust of the children, to include them in the regular activities of the program, and to talk with the mothers when they brought their children to school. She said that through gestures and human nonverbal kindness, she thought the children were feeling safe. But they seldom played with children other than those from their group of Somali friends and seemed to be picking up very little English. Hence other than by just observing them, she could not learn about their family stories, their interests, or their needs. She tried to ask for help from the mothers, but the language barrier prevented almost any communication.

One Sunday evening, in desperation about what she was going to provide for the children on Monday, she took a large garbage bag and went to her own children and her neighbors' children to ask for donations—of stuffed animal toys. The next morning, she entered the classroom and after greeting the children, she dumped the contents of the bag on the rug in the middle of the room.

There happened to be a stuffed camel among the animals, and the Somali kids jumped on it and started talking animatedly about the animal. Some went immediately to the sand table and started making what seemed to be a desert scene with dunes and troughs for water for animals and tents for the tiny plastic "people." Some other children went immediately to the art center and began drawing camels and their own versions of camel activities. Others went to the house center and began using the props available to prepare for some sort of feast.

The teacher was thrilled and immediately was able to ascertain from the children, with small bits of information from the mothers, that camels had been an integral part of these families' lives. They used camels for transportation, they raised them carefully and of course, became very attached to their family animals as American children do to house pets. They used the products of the camel for cooking and other life-maintaining needs. A study of camels, the teacher reported, ensued for at least 6 weeks. The events were important for the Somali children in that they could become the "experts" and teach the teachers and other children. In this method of teaching, they began to learn and use more English (and Spanish) in order to get their messages across. The information also was invaluable to the other children and the staff.

So What?

Dr Winsome Gordon of UNESCO spoke at an international Early Childhood Education conference in March 2002. She told of countless school visits to schools in many African and Asian countries. She spoke of the 6- and 7-year-old children coming to school with much real-world knowledge and then being handed a watered-down, inappropriate curriculum that treats them as immature innocents. She called for early years educators to acknowledge the experiences these students come to school with. Maybe it is the experience of caring for younger and elder family members, the experience of daily shopping and negotiating the family's food supply, the experience of cooperative survival in a refugee camp, the experience of survival during or cleaning up after war. While these students, of course, need the knowledge and skills taught in school curricula, their "funds of knowledge" (Moll, 1994) must be recognized and built upon. To begin to understand a culture, teachers must study its folk tales, legends, history, and the current culture of a group of people. It is not adequate to study only the ancestry of a culture and ignore how that culture has evolved through the ages.

Another student who participated in the Madlenka storybook and follow-up activity visited her grandmother's neighborhood and learned from some experts. She wrote about her experience:

> I visited two hair-braiding shops in Brooklyn. I gathered some information and background knowledge. I thought about history, science, much and art, building community and relationships. Not only did I learn more about people, their customs, and history, I learned more about myself as well. I was born here in America, I am of African descent with roots from the West Indies. In school I never really learned much about multiculturalism. I never learned a great deal about African civilizations, art, history, etc. African/black history is in fact very important in American history . . . Schools help shape one's identity and culture and vice versa. Self-concept has everything to do with how one views her/himself and how others view that person. How can one possibly go through life with marginal knowledge or misrepresentation of who they are and where they come from?

Schooling in these times is an experience in dissonance between two forces: on the one hand, a trend toward a global homogenization that brings people and countries closer than ever, and, on the other hand, the affirmation of what is specific and particular. This tension obviously has implications for decisions about teaching: are we concerned with the education of the citizen for a globally patterned culture or for a particular cultural identity? Why are we not concerned with both?

Mary Oliver (1984) asks in one of her poems, "Tell me, what is it you plan to do/with your one wild and precious life?" (p. 22). It is a challenge to us as learners and as educators. None of us lives or works in isolation and we draw our inspiration from a world community. A friend says with commitment that "The classroom is not an entity removed from the global moment, but is an expression of it" (Shome and Hegde, 2002, p. 184). There is hope.

References

Moll, L. (1994). Funds of Knowledge: A Look at Luis Moll's Research into Hidden Family Resources. *CITYSCHOOLS*, **1**, 1, pp. 19–21.

Oliver, M. (1984). *American Primitive*. New York: Little Brown.

Rummel, M. K. and E. P. Quintero (1997). *Teachers' Reading/Teachers' Lives*. Albany, NY: SUNY Press.

Shome, R. and R. S. Hegde (2002). Culture, Communication, and the Challenge of Globalization. *Critical Studies in Media Communication*, **19**, 2, pp. 172–189.

Chapter 12

Hiding in the Bathroom: The Educational Struggle of Marginalized Students

Danny Walsh

> *We were reading three books and they all dealt with houses. We were reading Vladimir Nabakov's Speak Memory, Isak Dinesen's Out of Africa, and Gaston Bachelard's Poetics of Space. When we got to the third book I was terribly confused and I couldn't make sense of what they were talking about. So I thought, it must be because I'm not smart enough. So I'll go to class and I won't say anything. But then it suddenly occurred to me when they started talking about the attic that they weren't talking about my house. We didn't have an attic in our house. You don't usually have an attic when you live in a third floor front.... Then I thought about all the books I ever had, all the way back to Dick and Jane and Sally and Spot. We never talked about my house. It was a horrific moment....*
> *I remember going home and getting so frightened that at that moment I think I could have given up my education. I felt I don't belong here.*
> Sandra Cisneros as cited in Menkart (1993, p. 4)

While researching multicultural education as a graduate student about a decade ago, I stumbled upon the above story recounted by Sandra Cisneros and felt profound reverberations with my own life; reverberations that compelled me to investigate the impact schooling has had and continues to have on me. Since reading Cisneros's words, I have engaged in a continuous peeling back of often resistant layers of my own history and identity to reveal my narrative and the disjunction I have experienced between schooling and my lived world. Above all, Cisneros's narrative brilliantly captures the feelings of those who have been rendered invisible and voiceless in schools through traditional curricula and teaching and learning processes. It is this voicelessness and invisibility, as imposed by the dominant culture and its grand narratives (the stories of the way that our lives should be) that often cause the psychological detachment that subsequently results in "poor performance" in school. Deemphasizing, dismissing, and/or ignorance of the complex connection between schooling and identity cause society to focus more readily—and simplistically—on the attributes of individual students that preclude them from academic success rather than analyzing the manner in which the power embedded in the teaching and learning processes renders some students more voiceless and invisible, and more susceptible to failure, than others.

Visualizing the above scenario, I cannot help but imagine that Cisneros's well-intentioned teacher wanted to expose her to "the great works," but ultimately taught her more through what was absent and implied in the course syllabus than about anything related to the metaphorical representation of houses in literature. As a poor, young person of Mexican descent growing up in Chicago, Cisneros may have never seen herself represented in school. Like many people of color, poor and working-class people, women, immigrants, and queers (to name a few) subject to the dominant culture's metanarrative of stable, middle class, white, heterosexual, and patriarchal homes, Cisneros felt confused, horrified, and even unintelligent because of the disjunction among her everyday life and curricula and pedagogy. Despite obvious divergence in our positionalities, I, as a gay person from a large working-class family, have also been and continue to be subject to such disjunction. With few exceptions have I ever felt truly represented in schooling. However, it has been during those moments of representation that I remember feeling most intellectually engaged and developing the most profound insights.

It is important to note here that I am not claiming the discrimination associated with racism and sexism experienced by women of color like Cisneros. Nor will I engage in the "oppression game." The aspects of my positionality that delineate me as "other" may not be immediately apparent; in addition to "otherness," I, as a white male, also occupy positions of power and privilege. Those who engage in the oppression game would surely locate Cisneros's positionality higher on the oppression scale; however, such rank ordering easily dismisses genuine feelings of disempowerment and exclusion and hinders movement toward solidarity on various fronts, a solidarity that "allows us to overcome impediments to self-direction together" (Kincheloe, 2001, p. 123).

With Cisneros's houses serving a springboard, I embark on an exploration of how my own school/life disjunction has impacted my understanding of knowledge (the epistemological), educational processes (the pedagogical), and what it means to be human (the ontological). In the words of Maxine Greene (1995), in an attempt to determine what my relationship is to some idea of the good, I "must inescapably understand [my] life in narrative form, as a 'quest' (Taylor, 1989, p. 52)" (p. 1). My narrative quest, as white gay man, as teacher, as working class, as brother and son, as New Yorker, attempts to reveal the reasons I seek out dangerous memories and a deeper understanding of the relationship between knowledge and power; appreciate different points of view and tolerate chaos, ambiguity, and not finding the right answers; and seem to pay attention to "the noise" in the background. Additionally, I hope to uncover why I have few memories of believing in the discourse of harmony, that we all bleed the same color blood and this blood automatically serves as the tie that binds. How has a history of feeling different, excluded, and voiceless in schools and within the dominant culture, yet not always possessing the language or conceptual framework to express such exclusion, contributed to my current state of being in the world?

It is not my intention to present myself as a self-actualized person who does not continually strive to hear "the noise," for example. However, there may be something about my location in the "web of reality" that predisposes me to the epistemological, ontological, and educational view briefly described above. Francisco Varela (1992) asserts, in his post-Cartesian view of cognitive science that challenges the traditional input-processing-output model, that knowledge is essentially about this situatedness; "and that the uniqueness of knowledge, its historicity and context is not a 'noise' concealing an abstract configuration in its true essence" (p. 7). The situatedness, the context, the historicity, the "noise" is the essence. My "moving, touching, breathing, and eating" (p. 8) in the world constitutes my being; I am not simply recovering a pregiven world but building knowledge from my microworld and microidentity (pp. 17–18). The world is not simply an "out there" thing waiting to be discovered; rather, the world is a construction of our complex interactions with it based upon the standpoint from which we view it.

In this endeavor to research my own life, I engage the emerging qualitative practice of critical hermeneutics, that is, "the art . . . [of] grappling with the text to be understood, telling its story in relation to its contextual dynamics and other texts, first to . . . [myself] and then to public audiences" (Kincheloe, 2001, p. 300). Criticality (the search for the always and already inscribed power) enters the scene as I strive to expose the power relations that have constructed my being. Such a methodology finds a partnership with the cognitive theory described above in that the architecture of the brain most likely supports an operation in which signals move back and forth, gradually becoming more coherent and thereby constituting the microworld, the world as we live it (Varela, 1992, pp. 48–49). Moving back and forth among texts while searching for inscribed power serves as the foundation of my narrative. Furthermore, I am attempting to understand the essence of a phenomenon, that of the disjunction between my everyday life and knowledge production in school and society. As with Sandra Cisneros, I rarely, if ever, saw my sometimes violent, almost-never-clean, crowded, financially insecure house represented in school. In this exploration, my narrative ". . . is not unlike an artistic endeavor, a creative attempt to somehow capture a certain phenomenon of life in a linguistic description that is both holistic and analytical, evocative and precise, unique and universal, powerful and sensitive" (van Manen, 1990, p. 39).

Maxine Greene (1995) utilizes the arts, particularly literature and painting, to recover and remember. Here, I follow suit defining texts broadly by including lived experiences. Greene believes, as do I, that "a reflective grasp of our life stories and of our ongoing quests, that reaches beyond where we have been, depends on our ability to remember things past" and we make sense of these things based upon where we stand in the web of reality (p. 20). My recovering and remembering include "texts" of experiences in school as both a teacher and a student; literature and other forms of artistic expression; and memories of my large working-class family, particularly my mother. Like Greene, "I cannot truly say [that this is] 'my life story.' That would imply that, spiderlike, I have somehow spun a web solely from the stuff of my own being, . . . I am not so 'individual' that I can claim to be free from the shaping influence of contexts. Nor can I forget that, conscious as I have tried to be, I have lived inside a whole variety of ideologies and discursive practices, in spite of trying— through resistance and critique—to liberate myself" (Greene, 1995, p. 74). Ultimately, this narrative quest is a form of liberation through self-disclosure and an effort to struggle for the transformation of schools through the constant work of transformation of self. I searched my story for epiphanies—life experiences that radically altered the meaning I give to myself—and discovered that my awakenings were often associated with interruptions of aspects of the dominant ideology that I have internalized, particularly those concerning race, class, gender, and sexual orientation.

Conscious examination of my own ideology, indeed searching for a sense of freedom, changes my way of being with others and thus teaching and researching. I have grown to welcome painful and dangerous memories, for such memories are not harmful as they remove me from the inertia and safety of the status quo and permit greater insight into "otherness." My quest for this insight begins with my mother. As with many re-memory narratives, I jump back and forth in time, at times mixing past events with present knowledge. Additionally, this narrative quest was not clearly mapped and inevitably led me down unforeseen paths. These "crooked paths," in contrast to the "straight roads of rationality" and "superhighways of positivism," allow me "to explore particularity, intuition, emotion, rage, cognition, desire, interpretation, experience, positionality, passion, social theory and knowledge in relation to one another" (Kincheloe, 2003, p. 195).

As cancer ravaged and eventually overcame my mother's body, images of her working herself to death crept into my mind. Today, several years after her burial, I am overwhelmingly convinced that, in addition to some environmental factors, my mother, as a working-class woman of her historical and cultural era, was relegated to a class and gender positionality that contributed to her premature death. An analysis of the gender, patriarchy, and class issues in my family not only helps to shed light on my dying mother, but also illuminates my epistemological, ontological, and pedagogical orientations. Essentially, memories of my working-class family and schooling experienced through this location write my counternarrative to the ". . . coercive and deforming effect of the dominant culture's official story, the metanarrative of secure suburban family life" (Greene, 1995, p. 164). In addition to Cisneros's houses, memories of my mother serve as the wellspring for interpreting my class location and how my feelings of socioeconomic "otherness" due to the devaluation of my lived experiences have consequently resulted in a search for windows into the worlds—the counternarratives—of "others."

In memory of my mother, my sister read the following at her funeral service: "I think back on my mom working 12-hour shifts, 7pm to 7am, and then coming home in the morning to make Christmas and Easter so special for all of us. She had to have been exhausted, but she never complained once" (personal communication, 2004). I believe that I am primarily aware of class and gender issues in my family and society because of the knowledge imparted by experiences reflected in my sister's writing. When I was about 12 years old, my mother, after raising nine children, returned to work out of economic necessity. Like many women who left home to return to work, my mother suffered double exploitation as she returned to a "feminized" job that required nurturing and taking care of others' bodies.

It was no accident that my mother was a nurse; as a woman first entering a profession at the end of the 1950s there were few alternatives. She would spend the last 20 years of her life working three or four 12-h shifts per week and doing odd jobs to supplement this income while continuing to manage a physically and emotionally demanding household. This double exploitation contributed to my mother's death as she cared for others, at work and home, but not herself. "As professional healers, midwives, and nurses, women have been repositories for and developers of knowledge about every one else's bodies" (Harding, 1998, p. 106). The surgeon general did not issue a health warning regarding the self-sacrifice associated with traditional notions of femininity (Kincheloe and Steinberg, 1997, p. 147) and if (s)he had I do not think my mother would have heard. When my mother needed a hysterectomy about five years before her cancer was diagnosed, she insisted, unsuccessfully, that she take the bus to the hospital so that others would not have to disrupt their lives to assist her. Later, as she suffered through chemotherapy, discussion of her illness was not an option, so much so that she never really came to terms with the fact that she might die.

As with many other women, my mother's reentering the workforce at the time of declining Western economies was accompanied by little change in the social dynamics and responsibilities at home (Weis, 1987). Here, patriarchy enters the picture as my father possessed, and my brothers and I inherited, the ". . . power men gain by birthright to define reality and enjoy the rewards by way of their domination of subordinates" (as cited in Kincheloe and Steinberg, 1997, p. 138). In the house of my youth, this power manifested itself in my father having few domestic obligations and more opportunities for relaxation at home. Cynthia Cockburn and Susan Ormerod (1993) have argued that while men view the home as a site of personal relations and recuperation from work, working women see it primarily as a site of domestic labor. In retrospect, my father was the victim of "male self-identity in patriarchal societies . . . grounded on the repression of affect, the disruption of connection. . . . [and the] lack of interpersonal connection . . . creating severe social dysfunctionality, especially in the areas of family, child care and women's issues" (Kincheloe and Steinberg, 1997, p. 138). I clearly remember arguing forcefully with my mother about my father's role in our house and saying that the only time I saw him help with domestic chores was when she was in the hospital having a baby. Moreover, I rarely saw or felt affection from him. The results of this patriarchy led two of my brothers, on separate occasions, without previously discussing the issue, to confront my father and blame him for my mother's death. I was not the only one to see my mother working herself to death, nor to recognize how patriarchy contributed to it.

Also at work here is the gendered separation of public and private domains in which the public world of work clearly takes precedence over the feminized home. Patriarchal ideology devalued the work my mother performed at home; in fact, it was not considered work at all. Christa Wolf (1989) "lists to herself the activities which the men of science and technology presumably do not pursue or which, if forced upon them, they would consider a waste of time: Changing a baby's diaper. Cooking, shopping with a child in one's arm or in the baby carriage. Doing the laundry, hanging it up to dry, taking it down, . . . Dusting. Sewing . . . Doing the dishes . . . Singing songs" (p. 31). I remember helping my mother with many of these waste-of-time tasks. This devaluation of my mother's private and public life ultimately contributed to her declining health. My mother saw herself and others saw her, using a naturalistic metaphor, as "endlessly bountiful . . . [and] in medical discourses, women's bodies have been modeled on a factory; when the machinery no longer functions, as in menopause, the factory has lost all of its value" (Harding, 1998, p. 100). Was it an accident that my mother had both a hysterectomy and breast removed in the last five years of her life?

I have few recollections of discussion of my mother's or father's work life. What I do remember is that my mother, being in a shortage area, enjoyed greater job stability and became better compensated for her work after years of experience, eventually almost doubling my father's salary. My father, as an airline mechanic, was subject to the economic tides of that industry. Before losing work completely with the decline of Pan American Airlines, the only company he had ever worked for, he was laid off during times of crisis and I remember him delivering newspapers and driving a school bus when it was unclear how long the layoff would last. I imagine his traditional notions of masculinity felt very threatened by job insecurity, possessing fewer opportunities for economic advancement than my mother, and the overall indignities associated with an unstable worklife. "One of the few domains of male workers' lives over which they can exert power and exercise control involves their relationships with women. . . . Scarred by the indignity of the workplace, many such husbands seek to re-establish their dignity through the domination of their wives in the domestic sphere" (Kincheloe and Steinberg, 1997, pp. 155–156). To this I would also add the domination of children. The physical abuse and emotional neglect inflicted by my father clearly had connections to patriarchal notions that rendered violence against women and children acceptable in the private domain. Enter schooling, the dominant culture, and the disjunction between home and academic life.

Joe Kincheloe and Shirley Steinberg (1997) define pedagogy as ". . . the production of identity—the way we learn to see ourselves in relation to the world" (p. 27), and cultural pedagogy as "the way individuals receive dominant representations and encoding of the world—are they assimilated, internalized, resisted, or transformed?" (p. 87). Such definitions depart dramatically from pedagogy's etymological roots that would more simply define it as the leading of children. A more complex interpretation of pedagogy sanctions the role that power plays in, and necessarily broadens the sources of, identity formation. For Henry Giroux (1997), pedagogy "involves the production and transmission of knowledge, the construction of subjectivity, and the learning of values and beliefs . . . Examining how people learn, make emotional investments, and negotiate the world around them, pedagogy is central to any discipline that studies educational and cultural processes vis-à-vis the making of meaning" (p. xiii). Cultural imperialism, a pedagogical process securely in place in schools, serves the normative function of centering one social group's experience while marginalizing and deeming deficient those groups located on the periphery.

This process also makes the marginalized groups' perspective invisible, stereotypes them, and designates them the other. With this complex pedagogical process in mind, I received innumerable messages about my family life, including the value of how we spoke, what we ate, and how we spent leisure time. I spent much of my adolescence separating myself from family because of these messages. Also, as part of the cultural imperialism process mentioned above, ". . . family values become whatever best reflects the preferred familial structure of the dominant group; marginalized family structures, no matter how well they might work for those living within them are deemed pathological by the power bloc" (Kincheloe and Steinberg, 1997, p. 85). Despite its limitations, there were times when my family structure worked; we were cared for and loved in many ways. However, I would rarely see the worldview emanating from my family and everyday life represented in school structures or curricula.

Although the disjunction between my school and family life was always present, it became most visible to me when I was 9 years old and we moved from urban Catholic schools to suburban public schools. Here, issues of class and race were more pronounced as the segregation of a small Long Island city made me more conscious of hostility toward whites and race and class distinction based upon imposed physical boundaries—very different from my experience in a racially diverse working-class neighborhood in Queens. We moved into the oceanfront city at a time when many storefronts on its main street were boarded up, with my father's realtor friend promising the place would take off in the next ten years. It was the late 1970s, and the city's school system seemed to still be feeling the impact of its integration efforts. The four elementary schools were located in distinct racial and class neighborhoods and we would be bussed to the school in the white working-class section from our house located just a few blocks from the city's housing projects.

Our school bus was a site of profound racial contestation among 5- to 10-year olds. The bus first passed through the housing projects picking up mostly black children who would, not incidentally I believe, occupy the back of the bus. We, noticing the segregation on the bus as it proceeded west, sat up front. Occasionally these self-imposed boundaries would be crossed and violence erupted with chants of "a fight, a fight, a nigger and white, if the _____ (nigger or white, depending upon who initiated the chant) don't win then we all jump in." This was my initiation into race and class consciousness—the political conflict between poor blacks and working-class whites embodied in children riding a yellow school bus.

This move from Queens to Long Island also brought with it a fourth-grade IQ test that identified me as "gifted." From this undeniably ability-grouped third and fourth grade class, I would be pulled out and bussed once a week to a district-wide enrichment program administered by Teachers College. The black students who occupied the seats at the back of the bus were in struggling readers' groups and pulled out for support services. In this program, I met for the first time my peers who attended the two elementary schools at the eastern, more affluent end of town. This would actually serve as a preview to what I would encounter throughout junior and senior high school where I would even more markedly feel the difference between the haves and the have nots as I was exposed to economic privilege and the academic success that often accompanies it.

While writing this, I suddenly remembered a few incidents during which I resisted this gifted program and thereby an academically talented identification: I clearly picture myself leaving my fifth-grade classroom under the guise of catching the bus waiting to deliver me to the program, yet I hid in the bathroom until I was sure the bus had left. A few years later I was threatened with removal from the program if I did not start showing some initiative. I was experiencing conflict and political struggle vis-à-vis my identity formation. I believe I was at some level attempting to emancipate myself from a situation in which I felt devalued and invisible, a situation in which everyone seemed to have more and therefore be more capable. My resistance, some of which would manifest itself in other ways throughout high school, also served to limit my access to knowledge that might elevate me from the working class. However, I was striving to protect my working-class worldview from a school that viewed me as lacking class and proper breeding.

If a random person on the street could state to my mother, surrounded by her many young children, "Don't you believe in abortion," it is not difficult to imagine the ideological constructions that influence how a kid from a large, Irish Catholic family might subconsciously be treated in school. "Students from subjugated groups typically feel that they are not a part of the school community, that they don't possess the secret knowledge that will let them into that club" (Kincheloe and Steinberg, 1997, p. 134). I was not part of that club; I had vastly different everyday experiences from the majority of my peers and I did not have the resources or academic support to participate in the social or intellectual networks. Like other marginalized students, I eventually concluded that schooling rewards those who already possess the dominant forms of cultural capital and I had to figure out how to negotiate this realization.

The dominant culture erased race and class from my schooling. It erased class through the four myths of equal opportunity, meritocracy, equality as conformity, and power neutrality in order to suppress any protestation to the maldistribution of wealth in society (Kincheloe and Steinberg, 1997, pp. 118–119). Like other dangerous topics, race and class are banished from the curriculum because of their divisive and demoralizing effects—the inclusion of racism and classism as an object of school study, would provide nonwhite and poor students with an excuse not to try (Kincheloe and Steinberg, 1997, p. 205). Moreover, a history of race and class conflict, domination and subordination, and cultural struggle might be revealed. Schools are rarely seen as appropriate sites for such revelation. Here, Giroux (1997) would contend that it is the culture of positivism that excludes such topics from the curriculum. Positivist educational practices emphasize objective, rational knowledge that separates the knower from the known thereby rendering it value free and neutral. Positivism presents such knowledge as "suprahistorical" and "supracultural." Within such an epistemological orientation, people are taught to believe that all things exist in isolation and problems are therefore detached from the social and political forces that give them meaning (pp. 11–13).

It was interpersonal difficulties and attitudes that caused violence to erupt on the yellow school bus and my family was to blame for what it lacked but it seemed that everyone else had. Moreover, such positivistic educational practice "excludes the role of values, feeling, and subjectively defined meanings in its paradigms" (Giroux, 1997, p. 19). The opportunity to discuss, literally and metaphorically speaking, my working-class family and school bus political struggles between working-class whites and poor blacks would not present itself. Discussion of less than harmonious relationships are dangerous and therefore disallowed. I was subject to ideologies that "generally fail to engage the politics of voice and representation—the forms of narrative and dialogue—around which students make sense of their lives and schools" (Giroux, 1997, p. 120). I would have to find a way to make sense and meaning and to negotiate the conflict between identification as both a member of the working class and an academic achiever. A positivistic educational system would never allow me a voice.

"Many of the alienated or marginalized are made to feel distrustful of their own voices, their own ways of making sense, yet they are not provided alternatives that allow them to tell their stories or shape their narratives or ground new learning in what they already know" (Greene, 1995, pp. 110–111). Like Greene, I felt alienated yet found a voice through texts that reveal the experiences of the "other." I particularly remember feeling a developing voice in a tenth-grade high school course entitled "Literature and Social Problems," an elective course taught by Ms. Vitale in which we not only read alternatives to the Western canon but also experienced an accompanying change in teaching style. While reading novels like *Native Son* by Richard Wright and discussing topics like environmental conservation and teenage suicide, we sat in a semicircle and actually spoke to other students. It is somewhat ironic, but not surprising, that I would have to diverge from my advanced placement track in order to more profoundly connect to academic life.

Later, I often wondered why it was that I could so clearly remember the horror of Bigger Thomas's transgression in *Native Son* and our discussions around the race and class issues presented in the novel. On what grounds could I possibly connect to a young black man living in abject poverty?

> Words like "nobodyness" and metaphors like "invisibility" were invented by black writers like James Baldwin and Ralph Ellison to describe their existence—helpless and unseen under the iron, an iron often moved back and forth by righteous men and women. I am not suggesting that I could claim kindred suffering, but at the same time . . . new meanings have fed into my own past shudderings and fears, and my present encounters with power, with force, with irrational pieties, with irons are extended somehow, and they are grounded in my lived world, my first landscapes, my "rememory." (Greene, 1995, p. 82)

Yes, there were outside forces at work, "the iron," the dominant ideology, that prevented me from seeing my own oppression and the oppression of others. Other people and their stories did matter and consequently my story mattered. I was angry, yet relieved. I encountered ". . . a pedagogy in which occurs a critical questioning of the omissions and tensions that exist between the master narratives and hegemonic discourses that make up the official curriculum and the self-representations of subordinate groups as they might appear in 'forgotten' or erased histories, texts, memories, experiences, and community narratives" (Giroux, 1997, pp. 156–157). In retrospect, it was no accident that I chose a class entitled "Literature and Social Problems"; I was on a search for lost narratives, particularly my own. I was "awestruck by truth-telling, motivated by the desire to get the whole story and moved by the vision of what education can be" (Kincheloe and Steinberg, 1997, p. 130). Much of the truth-telling in this class challenged notions of racial equity and the discourse of harmony. I was intellectually and emotionally intrigued by the subtlety with which racism manifests itself.

Kincheloe and Steinberg (1997) refer to these mutated, no longer overt forms of racism as crypto-racism, that is, racism that is ". . . concealed, secret, and not visible to the naked eye. Often, such a racist form employs thinly veiled racialized code words that evoke images of white superiority" (p. 195). To this day, I am still somewhat taken aback, perhaps naively, by the "codes" used by teachers when speaking about students of color. Once I was discussing with another teacher some challenging behavior problems that I was having with a student, with the intent of problem solving, and she ended the conversation by stating, "Some of them are straight from the jungle." Being new in the school building, I did not know how to react. I was shocked. I was complicit and allowed that talk to pass uninterrupted. She did not even realize the connotations of her words. Perhaps a more subtle incident involved the same teacher who thought that the solution to many students' academic deficiencies was "good dinner time conversation." I recently read a response to this that asked what Abner Louima's (a man sodomized by the New York City Police Department) family would discuss around their dinner table that might be incorporated into a literacy lesson. Again, this teacher failed to hear the biases in her words. I often wondered what this woman's classroom, filled with mostly poor students of color, felt like to them. "As it engages in the language of denial, crypto-racism aligns itself with notions of: (a) a common culture, i.e. all solid citizens hold particular, sacred values and understanding (implicit in the assertion is that many non-whites do not share these values and understanding)" (Kincheloe and Steinberg, 1997, p. 195). These students did not know how to respond to authority, were therefore less than human, did not have white middle-class cultural and political conversations around the dinner table, and they were therefore ignorant. With such an ethos established by a teacher, is it any wonder that students "act out"?

I was also not a holder of the sacred values. I acted out. I protested. I resisted. As briefly mentioned above, my acts of resisting academically talented identification, and subsequently school, emanated from the conflict I felt between dual identification as working class and "gifted." As also mentioned, I attempted to separate myself from my family because I perceived I needed to be more like those who succeeded in school, the economically privileged, so that I could become a member of the club. "School is sometimes like a jealous lover who demands that marginalized students must choose between their peers or school—if school is chosen then one must give up one's culture and adopt the identity of a school achiever" (Kincheloe and Steinberg, 1997, p. 127). I alternately chose school and my family. I achieved and did not achieve in school. I achieved to say "hey, look at me, I'm doin' this anyway" and I did not achieve because I felt disconnected and hopeless about future possibilities.

In many ways, I was a compliant and well-behaved student. Perhaps this is why I was allowed to pass by the gatekeepers who would have been more inclined to chose a more privileged peer for participation in any enrichment programs. Perhaps I was admitted into the gifted program without meeting its criteria because a teacher thought that a student from a large working-class family deserved a chance. Obviously, at this point, this is mere speculation on my part. Whatever caused me to hide in the bathroom so as not to participate in a gifted program, at least on that one day, was the result of internal conflict and a sense of "otherness." Olugbemiro Jegede (1999) uses collateral learning theory to explain how non-Western learners cope with learning science in a classroom that is hostile to their indigenous knowledge. Ultimately, they learn to separate indigenous and Western ways of knowing as they "grapple with the need to resolve understanding from two domains on an everyday basis" (p. 131). Collateral learning reminds me of a recent speaker at the CUNY Graduate Center who discussed the distance, epistemological and emotional, that students "travel" to arrive at school. I sometimes felt as if I had walked miles to get to school.

School was an emotionally draining experience from which I often could not wait to escape. In junior high and high school, I resisted social identification as an academic achiever by smoking cigarettes, drinking alcohol, and later using drugs because I thought these things might inscribe an alternate identity. I also resisted school knowledge.

"Marginalized student resistance to mainstream norms often expresses itself in terms of a cultivated ignorance of information deemed important by the so called 'cultured' " (Kincheloe and Steinberg, 1997, p. 100). I remember feeling more and more alienated from school as the years progressed and spending more time working a part-time job than doing homework, primarily to have things and to be able to do things. In school, of all places, I was developing anti-intellectual tendencies and becoming an accomplice to my own subjugation (Benjamin, 1977, p. 22, as cited in Giroux, 1997, p. 57).

The removed-from-my-everyday "advanced" knowledge and "higher" culture of secondary school became more foreign to me. Master narratives seemed to become more commonplace. On more than one occasion, Greene (1995) refers to Toni Morrison's character Pecola Breedlove in *The Bluest Eye* and the "way in which [she] is destroyed by two of the dominant culture's master narratives: the *Dick and Jane* readers and the mythic image made of Shirley Temple with her blue eyes" (p. 118). As with Sandra Cisneros, we see the destructiveness of the dismissing of the "other's" narrative—other than white, middle-class, stable homes. Again, I am not claiming the impact of racism and sexism, but I undoubtedly felt the jolt as the cultural and socioeconomic "other."

Here I visit the notion of the dominant culture's determination of what constitutes "high" versus "low" or popular culture, and, as mentioned above, the alienation I have felt from dominant cultural narratives in literature and other forms of artistic expression. There was "classic" literature that spoke to me; however, it seemed that the "high" cultural world of classical music, painting, sculpture, museums, and the theater as well as the food and drink that accompanied such activities was the secret knowledge of others. It was a world into which I was not invited, with the exception of an occasional school trip, and I feel the repercussions even today as evidenced through my discomfort in museums, galleries, and yes, even restaurants. The dispositions and preferences evident in my sense of appropriateness and validity of taste for cultural goods, which not only operate at the level of everyday knowledge but are also inscribed on my body, might reveal me in these venues (du Gay et al., 1997, p. 97). My discomfort emanates from a fear of being found out, perhaps being viewed as "uncultured" or "unintelligent" because I would not recognize art work or food, or know how to talk about it, or know how to act in these settings.

In the words of Tillie Olsen (1978), this world of high culture was " 'dark with silences,' the 'unnatural silences' of women who worked too hard [like my mother] or were too embarrassed to express themselves (p. 6) and of others who did not have the words or had not mastered the proper way of knowing" (as cited in Greene, 1995, p. 160). The proper way of knowing was withheld from me. I have recently found solace in the "low" or popular culture of quilts stitched from rags by poor Alabaman African American women displayed at the Whitney Museum; the choreography of Arthur Aviles performed in a dance studio converted from an old warehouse in the South Bronx; drag queens working the stage in Tompkins Square Park; the Great Migration paintings of Jacob Lawrence; the poetry of Emanuel Xavier, a former queer hustler, who worked the West Side Highway piers; and the pulsating dance beats spun at Body & Soul, a now defunct club, sold to the highest bidder as Tribeca further gentrified, that was filled with gay men and lesbians of every color. Obviously, my attraction to such cultural productions is about much more than class status. Where are there spaces to discuss the pedagogical processes involved in such art forms?

Seeking out art and culture such as this is my postmodern critique, my challenge to the master narrative, and my opposition to the Western canon and the common culture it attempts to develop. "Postmodern thinkers challenge hierarchical structures of knowledge and power which promote 'experts' above the 'masses,' as they seek new ways of knowing that transcend empirically verified facts and 'reasonable,' linear arguments deployed in the quest for certainty" (Kincheloe and Steinberg, 1997, p. 38). A critique such as this is "the explosion of reification, a breaking through of mystifications and a recognition of how certain forms of ideology serve the logic of domination" and determine what is included and what is excluded (Giroux, 1997, p. 85). While many would consider painter Jacob Lawrence worthy of appreciation and investigation, few would attribute such status to a former queer hustler's poetry or to a queer choreographer who lives and performs in the economically depressed South Bronx. "Knowledge that emerges from and serves the purposes of the subjugated is often erased by making it appear dangerous and pathological to other citizens" (Semali and Kincheloe, 1999, p. 32). In fact, the people themselves who produce these knowledges are rendered dangerous and pathological by the dominant culture's ideological standards; the subjugated are portrayed as witches, heretics, demons, savages, and faggots, and are summarily dismissed.

Like Sandra Cisneros, despite being dismissed and feeling like I did not belong, I did not give up my education. I emerged from the bathroom that protected me from further insult and continued to attend the "gifted" program. Even as a doctoral student, I sometimes return to this metaphorical place of refuge when my vocabulary and experiences seem limited, when I miss the punchline of jokes with references to "high" culture, when I do not know what others around me seem to know, and when heterosexual assumptions render me invisible. Unlike many marginalized students, I have found consolation in the academy. With the help of teachers like Ms. Vitale in classes with titles like "Literature and Social Problems" as well as my own counter-cultural explorations, I have encountered voices that told stories both like and unlike mine, but always different from the dominant culture's narratives of what and how I/we should be. Schooling, as a reflection of these narratives, too often fails to recognize alternative ways of being in and seeing the world and subsequently fails students marginalized because of race, class, gender, and sexual orientation. My argument here is primarily one for the inclusion of dangerous memories in the curriculum for such memories centralize the periphery and provide schooling with the meaning so desperately needed and desired by marginalized students. We are all too familiar with what happens to students who do not find such meaning, who do not emerge from their place of hiding.

References

Cockburn, C. and S. Ormerod (1993). *Gender and Technology in the Making*. London: Sage.
du Gay, P., S. Hall, L. Janes, H. Mackay, and K. Negus (1997). *Doing Cultural Studies: The Story of the Sony Walkman*. London: Sage.
Giroux, H. (1997). *Pedagogy and the Politics of Hope: Theory, Culture, and Schooling*. Boulder, CO: Westview Press.
Greene, M. (1995). *Releasing the Imagination: Essays on Education, the Arts, and Social Change*. San Francisco, CA: Jossey-Bass.
Harding, S. (1998). *Is Science Multicultural?: Postcolonialisms, Feminisms, and Epistemologies*. Bloomington, IN: Indiana University Press.
Jegede, O. (1999). Science Education in Nonwestern Cultures: Towards a Theory of Collateral Learning. In L. Semali and J. Kincheloe (eds.), *What is Indigenous Knowledge?: Voices from the Academy*. New York and London: Falmer Press.
Kincheloe, J. (2001). *Getting beyond the Facts: Teaching Social Studies/Social Sciences in the Twenty-First Century*. New York: Peter Lang Publishing, Inc.
Kincheloe, J. (2003). *Teachers as Researchers: Qualitative Inquiry as a Path to Empowerment* (2nd edn). London and New York: RoutledgeFalmer.
Kincheloe, J. and S. Steinberg (1997). *Changing Multiculturalism*. Buckingham and Philadelphia: Open University Press.
Menkart, D. (1993). Multicultural Education: Strategies for Linguistically Diverse Schools and Classrooms. Washington, DC: National Clearinghouse for Bilingual Education (ERIC Document Reproduction Services No. ED 364 106).

Semali, L. and J. Kincheloe (eds.) (1999). *What is Indigenous Knowledge?: Voices from the Academy.* New York and London: Falmer Press.

van Manen, M. (1990). *Researching Lived Experience: Human Science for an Action Sensitive Pedagogy.* Albany, NY: State University of New York Press.

Varela, F. (1992). *Ethical Know-How: Action, Wisdom, and Cognition.* Stanford, CA: Stanford University Press.

Weis, L. (1987). High School Girls in a De-Industrializing Economy. In L. Weis (ed.), *Class, Race, and Gender in American Education.* Albany, NY: SUNY Press.

Wolf, C. (1989). *Accident: A Day's News* (H. Schwarzbauer and R. Fakrorian, trans.) New York: Farrar, Straus & Giroux.y

Chapter 13

When Ignorance and Deceit Come to Town: Preparing Yourself for the English-Only Movement's Assault on Your Public Schools

Pepi Leistyna

There is a fierce battle being waged over language policy and practice in public schools in the United States. Both proponents and opponents have focused their arguments on which language of instruction will linguistically diverse students learn best.

On the one hand, antibilingual advocates have argued that in order to promote effective nationwide communications and meet the demands of modern technology and the economy, the United States is compelled to use a linguistic standard. These political voices thus call for a mandatory English-only approach for all children in public schools throughout the country. At the forefront of this cause is Ron Unz the chairman of the national advocacy organization English for the Children, and the originator of California's Proposition 227, which in 1998 effectively outlawed bilingual education in that state. After a similar victory in Arizona in 2000, he also attempted to win over Colorado. However, a wealthy parent spent a mountain of her own personal money on a press campaign to convince the middle-class white population not to support Unz's initiative because if bilingual programs are dismantled, then "those kids will be in class with your kids." This well-funded, racist plea worked and Colorado voted no on the English-only referendum. Unz is currently focusing on New York and Oregon.

Unz demands that the United States replace bilingual education (which he describes as "a disastrous experiment") with a one-year

Structured English Immersion Program. As the English for the Children publicity pamphlet states

> Under this learning technique, youngsters not fluent in English are placed in a separate classroom in which they are taught English over a period of several months. Once they have become fluent in English, they are moved into regular classes.

When asked by a reporter, "Won't immigrant kids fall behind in other subjects besides English if they aren't taught in their own languages?," Unz (1997) replied, "The vast majority of the students involved [in linguistic transition] enter school when they're just 5 or 6, and at that age, it takes just a few months to learn English." Without reference to any specific theory or research to back up this claim, he asserts that

> human brains at a young age are designed or wired up for language acquisition. And that's what all the neurological science indicates. It's what every ordinary person in the world believes, but it's contrary to the theory of bilingual academics.

As no specific research literature is cited, it is unclear if Unz is attesting to the validity of Noam Chomsky's notion of the Language Acquisition Devise (LAD), or if he is laying claims to Steven Pinker's idea that language is a human instinct that is wired into the brain by evolution. Unz could also be referring to the most recent research that identifies FOXP2—a specific human gene that affects the brain circuitry that makes possible language and speech.

There is strong evidence to suggest that human beings are biologically predisposed with certain cognitive structures that facilitate language growth and logical thought. It is probable that, as Chomsky has argued, there is a language-specific organ of the mind that provides an in-house abstract blueprint known as universal grammar, against which language acquirers can test hypotheses and develop surface language syntax. If Unz is situating himself in this Innatist school of thought, he neglects to elaborate on the details of the theory that informs his political motivations toward English-only. The fact that he does not offer up recognition that such psycholinguistic tools do not predispose humans to knowledge, using language in social contexts, body language, literacy, or critical inquiry, is evidence of his lack of expertise in this area. English-only advocates do not detail their explanations of how children actually acquire language or develop literacy skills, or how to assess such growth. Unz's diatribes do not even differentiate speaking from reading and writing—two interrelated but very different abilities. To make the English-only political position vaguer, advocates, while staking their claims to mandatory English language instruction, offer no pedagogical detailing of such classrooms.

Worlds apart from Unz's claim that "It's what every ordinary person in the world believes, but is contrary to the theory of bilingual academics," I find that it is very easy to introduce Innatist concepts to language educators in my Applied Linguistics Graduate Studies/teacher education program, but very difficult to explain to the general population. Most people unknowingly subscribe to a behavioristic stimulus–response–reward explanation of language development. That is, they readily believe that children learn language by mimicking the people around them. In addition, in my experiences, the public easily falls prey to the logic that the more time on task, in other words, the more English you speak the more you will learn. Within this realm, there is little patience for the counterintuitive and research-backed logic that quality education in the first language facilitates the growth of the target language—that is, what is more understandable in the first language will be easier to make sense of in the second.

Many Republicans, and Democrats alike, embrace the national movement toward English-only language and literacy policies and practices. In Massachusetts, where Unz's initiative recently passed in a state-wide referendum, teachers are desperately looking for guidance in how to go about infusing this new approach to language learning, and teacher education programs are scrambling to restructure their bilingual education services in order to prepare teachers to work in sanctioned, Structured English Immersion classrooms.

The Assault on Massachusetts: The Spoils of Propaganda

In November 2002, after being bombarded with misinformation, the Massachusetts voters decided the fate of bilingual education in K-12 public schools. English for the Children came to Boston in order to work to replace Transitional Bilingual Education, a three-year program, with one year of Structured English Immersion. Unz, the outspoken leader of this anti-bilingual initiative, drafted the Massachusetts proposal. What are the claims made by Unz and his followers, and what facts contradict this agenda?

In Unz's view, linguistic-minority students require only one year of Structured Immersion in an English-only context in order to join native speakers in mainstream classes. However, research clearly shows that it can take children from five to seven years to become fluent and literate, able to learn sophisticated content in the second language, and thus able to handle the demands of standardized testing like the Massachusetts Comprehensive Assessment System (MCAS). This should come as no surprise as that is how long it took all of us in our first language experience. Effective immersion and bilingual programs take this fact into account and they work from the basic premise that if knowledge is comprehensible in the first language (e.g., the language of math), then it will be easier to understand in the second language. The catch-up process in bilingual education consequently includes grade-appropriate content in the native language while the English improves.

Countering the criticism that one year of Structured English Immersion and subsequent mainstreaming will lead to certain failure of so many students, Unz's promotional pamphlet states, "It [the organization] will NOT throw children who can't speak English into regular classes where they would have to 'sink or swim.' " So, rather than supporting bilingual education's simultaneous development of knowledge, language, and literacy skills, students will remain in a segregated and mixed-age holding tank in English where they will be served watered-down curricula in other content areas. It is ironic that support for stifling any real advancement in Math, Science, Social Studies, and the Language Arts comes from a movement that is supposedly seeking to make education more rigorous.

There is no defensible theory or body of research to support the claim that students need only one year (about 180 school days) to become fully fluent, literate, and capable of learning content in another language. Imagine yourself going to another country where you did not speak the language and pulling off this massive feat in such a short period of time. Those of us who have actually studied overseas can fully appreciate the time and support necessary for this process.

Regardless of Unz's rhetorical claims, the majority of students in California in Structured English Immersion did not achieve even intermediate fluency after one year. Take, for example, the Orange Unified School District that is so often used to support Unz's argument: after the first year, 6 out of 3,549 students were mainstreamed; more than half of the students were not ready for his specially designed classrooms. A more recent progress report in California reveals the extent of the disaster:

> In 2002–2003, it [Ron Unz's Structured English Immersion] failed at least 1,479,420 children who remained limited in English. Only 42 percent of California students whose English was limited in 1998, when Proposition 227 passed, have since been redesignated as fluent in English—five years later! (Crawford, 2003, p. 1)

With five years of watered-down content, rather than intensive subject-matter instruction in the primary language, these students will certainly be ill-prepared for high stakes standardized tests. In states like Massachusetts, students who do not pass the state's standardized test in high school will not graduate. Instead, they will be shown to the door and handed a Certificate of Attendance on their way out; that is, if they manage to stay in school under such conditions.

English-only advocates claim that bilingual education causes high dropout rates, especially among Latino/a students. The reality is that only a minority of Limited English Proficient (LEP) students around the country are in bilingual programs. The real issue at hand is that 75 percent of all linguistic-minority students reside in low-income, urban areas that have schools that are highly segregated and in rough shape. These students so often face harsh racist and material conditions, incessant harassment, segregated school activities, limited classroom materials, ill-prepared teachers, poorly designed and unenforced policies, and indifferent leadership that dramatically disrupt their personal, cultural, and academic lives. Unz disregards these political and ethical issues and simply calls for "structured mixing," when possible, of mainstream and Structured English Immersion students. Not only does his plan avoid confronting the discrimination that takes place in public educational institutions—that in fact leads to high dropout rates—but it is also unclear how such a strategy for integration is going to work in schools that are segregated because of economic/housing demographics.

Unz maintained that his legislative intent is only an extension of Latino/a parents' discontent with bilingual education. Contradicting this stance, according to the National Council of La Raza (NCLR), which is one of the nation's leading Latino/a advocacy groups, 80 percent of registered Latino/a voters across the country are in favor of bilingual education.

Unz told the people of Massachusetts that the standardized test scores of over 1 million students have improved in California as a result of Proposition 227. However, as correlation does not necessarily speak to causality, he failed to inform the public about the likelihood of test inflation caused by teaching to the test, special test preparation, selective testing in terms of who gets to participate, class size reductions, political and financial incentives for those administrators who comply, and eliminating most school activities that are unrelated to the test. In fact, the scores of all the students improved, including those from bilingual programs—what little remains of them. In some cases, students in bilingual education scored higher overall. What has not significantly decreased in states that have already adopted English-only programs is the gap between the LEP students and the mainstream. Of course, Unz said nothing of this during his Massachusetts campaign.

Unz insisted that linguistic-minority students are trapped in bilingual programs for most, if not all, of their public school careers—programs that he emphatically argues are supported by faulty educational theories rather than empirical data. Contrary to Unz's assertion, the average stay in bilingual classrooms in Massachusetts is 2.8 years. In exceptional programs in cities like Framingham and Cambridge, the average is 2.3 years. In addition, there are over 150 studies that show how, when properly implemented, including the native tongue is beneficial for linguistic, psychological, cognitive, and academic development.

Unlike Unz's one-size-fits-all approach, students from diverse age groups and backgrounds have different needs. When these needs are addressed, and parents are allowed to be part of the assessment process and are given an informed choice as to what is best for their children, the results are outstanding. But bilingual education's success or failure depends entirely on the people and institutions that bring such programs to life. Unz ignores successful examples such as Framingham's: in 2001, out of the 1,500 students that are in one of the five language programs that are offered, 92 percent of the third graders passed the state's standardized test (the MCAS) in English. The teachers responsible for such success, under the Unz initiative, can be sued and banned from working for five years if they use any language other than English in the classroom.

Unz calls his campaign English for the Children, but the evidence suggests that it will not achieve such benign ends. If we truly want to raise the English-speaking skills of nonnative speakers, we can surely find more reliable guidance than this draconian solution championed by a monolingual, multimillionaire with no children, and with no background in education or linguistics. But such guidance is difficult when set up against big money and corporate media. While Unz and his spokespersons got open access to newspapers and other media, it was very difficult to get alternative perspectives out to the public. Believe me, I tried! Even with a Masters and Doctorate in language and literacy from the Harvard Graduate School of Education, dozens of journal articles on language and education, five books with major publishing houses, years of experience teaching English as a second language and in a teacher education program in Applied Linguistics, I was virtually unable to put forth a counterargument to Unz's assault on bilingual education. After dozens of attempts with the major local newspapers, it was only the *Cambridge Chronicle* that took any interest in publishing one of my op-eds. Likewise, when major scholars like Jim Cummins and James Crawford came to the University of Massachusetts, Boston to speak on the issue, the press, though invited, was nowhere to be found. This was particularly telling and disturbing given that the *Boston Globe* is laterally right across the street from campus.

Playing with Patriotism, Demonizing Difference

Capitalizing on the public's general discontent with K-12 schools, proponents of English-only have worked tirelessly and effectively to scapegoat bilingual education, creating legal constraints on the daily lives of educators by ensuring that languages other than English (with the exception of "foreign language instruction") are stomped out of school life entirely. In an effort to do so, anti-bilingual forces have capitalized on public fears over national unity. The U.S. English Foundation, Inc. believes "that a shared language provides a cultural guidepost that we must maintain for the sake of our country's unity, prosperity and democracy . . ." The English for the Children pamphlet adds, "Children who leave school without knowing how to read English, write English, and speak English are injured for life economically and socially." Beyond the ethnocentric assumption that literacy and economic and technological effectiveness can only take place in English, proponents of the English-only movement assume that the fundamental reason that the country potentially faces internal turmoil is because of the failure and/or unwillingness of linguistic minorities to assimilate. Not only does this dehistoricized position presuppose that the country has at some point been united, but its ideologues also strategically say nothing about a system within which people are relegated, and not by choice, to live on the margins of economic, social, and political power.

The leading voices of English-only say virtually nothing about the socially sanctioned and systemic practices that discriminate against certain groups of people and that generate antagonistic social relations and economic exploitation and abuse. For example, there is no detailed critique of the recent anti-immigrant and anti-Affirmative Action conflicts and racial strife that is embodied in California's Propositions 187 and 209. While Unz proclaimed to be an opponent of Proposition 187, it was unclear if he simply needed an opposing platform to Pete Wilson's as they were competing for the governorship, or if he was in no hurry to rid his state of a cheap labor force—after all, he is the owner of a software company. Nor is there adequate public concern for, or media coverage of, the fact that schools throughout the country remain profoundly racially segregated. In addition, there is little to no discussion of the hidden agenda (or not so hidden in the case of Colorado) of conservative support for bilingual education that embraces improper implementation so as to ensure that racially subordinated children are segregated from privileged whites. Only when put on the spot, during an interview, does Unz himself admit to this reality:

> Although this is a touchy point, there does seem to be some anecdotal evidence that it's sometimes true. . . . Under this analysis, bilingual education represents mandatory racial segregation, which makes it even stranger that it's become part of liberal orthodoxy.

In the first half of his statement, Unz makes no effort to excoriate the racist implications of such actions; in fact, he shows no distress at all. The logic in the second half of his response implies that simply because racists work to misuse a program (which is not designed to be about segregation during the entire school day), then supporters should simply give up on it. That's the equivalent of saying, because corporations such as Enron, WorldCom, and Tyco abuse democracy, then we should all abandon such a governmental process.

Ironically, some anti-bilingual advocates, such as former president Ronald Reagan, insist that instruction in languages other than English is un-American. This paradoxical twist disregards that the Constitution of the United States protects linguistic pluralism, and that the U.S. Supreme Court's 1974 Lau v. Nichols decision was intended to protect the rights of linguistic minorities in public schools. States like Alaska and Oklahoma found English-only practices in government to be unconstitutional. It also seems more unpatriotic for a democracy to exclude (or mark as "foreign") languages that are now indigenous to the United States: the native tongues of Puerto Rico, Native America, Hawaiians, African Americans, and Mexican Americans.

It is also curious that so many mainstream politicians concerned with public education work so hard to eradicate multilingualism among racially and economically oppressed students, while simultaneously working to make certain that upper-middle class and wealthy youth are able to speak international languages. Multilingualism, which is embraced in all the finest private schools in the country, worldwide for that matter, is great for elite children but somehow bad for, and unpatriotic of, the poor.

In addition, as democracy and commonality are a contradiction in terms (i.e., democracy requires difference, participation, and dissent, rather than conformity through coercion), it is the proponents of English-only and common cultural literacy that in fact embrace social fragmentation. In other words, their academic canons and linguistic standards exclude by their very nature. From a self-professed opponent of bilingual education, Affirmative Action, multiculturalism, and multicultural education, and most other movements and policies that support a more participatory democracy, we hear statements from Unz (2000) such as:

> First and foremost, our public schools and educational institutions must be restored as engines of assimilation they once were. . . . In history and social studies classrooms, "multicultural education" is now widespread, placing an extreme and unrealistic emphasis on ethnic diversity instead of passing on the traditional knowledge of Western civilization, our Founding Fathers, and the Civil and World Wars . . . current public school curricula which glorify obscure ethnic figures at the expense of the giants of American history have no place in a melting pot framework. (Unz, 2000, pp. 3–4).

In a cover story in *Commentary* (1999), with the shock value title of "California and the End of White America," Unz is able to maintain the existing and balkanizing fear in many whites that they are being overrun, while at the same time scaring racially and ethnically diverse peoples with the "inevitability" that there will be white backlash against them in the form of "White Nationalism." He warns:

> Our political leaders should approach these ethnic issues by reaffirming America's traditional support for immigration, but couple that with a return to the assimilative policies which America has emphasized in the past. Otherwise, whites as a group will inevitably begin to display the same ethnic-minority-group politics as other minority groups, and this could break our nation. We face the choice of either supporting "the New American Melting Pot" or accepting "the Coming of White Nationalism." (Unz, 1999, p. 1)

What is particularly interesting about this rhetorical strategy that calls for "assimilative policies" is that the mainstream that supports U.S. English-only is not the least bit interested in the assimilation of racially subordinated groups into their neighborhoods, places of work, educational institutions, clubs, and communities, that is, in equal rights and universal access. As a person said to me today when he learned that I had to go home after the football game to finish writing this article on language policy and practice in the United States:

> Those people need to learn English and be taught only in English. You and that multicultural stuff . . . keep it up and they'll be living right next to you!

As has been the case historically, under a xenophobic climate clouded with anti-immigrant sentiments, the main concern of local folks is with "unwelcomed outsiders" taking over of jobs and affordable housing, and flooding public schools and other social services.

The harsh reality is that beyond the concocted hype about the usurping of quality employment by "outsiders," the job opportunities that are intended for migrant workers, the majority of immigrants, and the nation's own down-trodden, consist of manual labor, cleaning crews, the monotony of the assembly-line, and farm jobs that require little to no English—as with the Bracero Program (1942–1964) when more than 4 million Mexican farm laborers were "legally brought" into the United States to work the fields and orchards. These workers spoke little to no English, signed contracts that were controlled by independent farmers associations and the Farm Bureau, and were immediately put to work without an understanding of their rights. In 1964, when the Bracero Program was finally dismantled, the U.S. Department of Labor officer heading the operation, Lee G. Williams, described it as "legalized slavery."

There is a new scramble by big business and politicians, both Republicans and Democrats, to "legalize" undocumented workers. In response to a demand and shortage of low-wage, low-skill jobs, George W. Bush's administration is looking into another guest worker program. Thus, being pro-immigrant, as Unz claims "Nearly all the people involved in the effort [English-only] have a strong pro-immigrant background," does not necessarily mean being pro social justice.

As the founder and chairman of a Silicon Valley financial services software firm and the 1994 GOP nomination for governor of California, Unz's insistence that an English-only approach will ensure "better jobs for their [linguistic-minority children's] parents" does not seem to ring in solidarity with organized labors' concerns with the systematic exploitation of workers, both documented and undocumented. Simply shifting to a one-year sink or swim Structured English Immersion Program for what would now be "legal" workers (who by the way will not be going to school as they will be working long hours) will not eradicate the problems of economic abuse and subjugation. On the contrary, such conservative programs provide limited access to language and learning and prevent most linguistic-minority children from attaining academic fluency in either their native language or in English. A one-year Structured English Immersion Program is surely designed to fail in developing both fluency and literacy. Instead, these kinds of state and federal educational policies and practices reflect an implicit economic need to socialize immigrants and members of oppressed groups to fill necessary, but undesirable, low-status jobs.

The Colonial Language Trap

Taking away the native tongue, while never really giving access to the discourse of power, is a common practice in any colonial model of education. Such a deskilling process creates what I refer to as bridge people: people who are miseducated in a way that connects them to two worlds but works vigilantly to make certain that they belong to neither. This strategy effectively works to deny their access to the mainstream while simultaneously taking away any tools that can be used to build the cultural solidarity necessary to resist exploitation and transform society. It is thus clear that assimilationist agendas are really about segregation. Homi Bhabha's (1994) concepts of "ambivalence" and "mimicry" shed light on how the myth of assimilation works.

In the operations of colonial discourses, Bhabha (1994) theorized a process of identity construction that was built on a constant ideological pulling by a central force from contrary directions in which the "other" (the colonized) is positioned as both alien and yet knowable; that is, deviant and yet capable of being assimilated. In order to keep the colonial subject at a necessary distance—unable to participate in the rights of full citizenship, stereotypes are used to dehumanize the oppressed, while benevolence and kind gestures are superimposed to rehumanize them. To use a current example, Latino/as in the United States are represented as lazy, shiftless, violent, and unintelligent— dehumanized by the press as "illegal aliens" and "non-white hordes." The language of popular culture embraces more blatant racist language: "border rats," "wet backs," "spicks," and so on. A popular joke reveals this deeply embedded racism and callousness: "What do you call a Mexican with eight arms? I don't know, but you should see it pick tomatoes." And yet, these same people are simultaneously deemed by English-only advocates as worthy of good education, standard language skills, employment, and advancement. From Bhabha's perspective, it begins to make sense why conservative politicians and organizations such as the U.S. English Foundation, Inc. make claims to disseminating "a vehicle of opportunity [English] for new Americans."

As an essential part of this process of maintaining ambivalence, colonizers need members of the subordinated classes that can speak the dominant tongue, and express its values and beliefs as superior and benevolent "gifts." This is exemplified in the United States in the work of Richard Rodriguez, Dinesh D'Souza, and Jaime Escalante who served as the honorary chairman of Proposition 227. Bhabha refers to these agents as "mimic men," but adds (referring to the British in colonial India), "to be Anglicized is emphatically not to be

English" (p. 87); in the case of the United States—to be Americanized but never really accepted as American.

The position of flux that ambivalence invokes could lead to political resistance inside the ranks of the colonized. These "mimic men" (and women) can be a menace to the colonizers as they have access to the cultural capital and strategies used by the colonizer to maintain the material and symbolic system of oppression. As John McLeod (2000) explains, "Hearing their language returning through the mouths of the colonized, the colonizers are faced with the worrying threat of resemblance between colonizer and colonized" (p. 55). Unlike the bridge people described earlier, these forces of resistance that are able to effectively navigate both worlds can work to transform the inhumane symbolic and material conditions that so many people are forced to live in on a daily basis.

Not surprisingly, anti-bilingual proponents tell the public virtually nothing about that horrific material and symbolic conditions that so many children and young adults face in this discriminatory society and in the schools that reflect this larger social order; they are not told about the one-in-five children—one-in-four racially subordinated youth(s), who grow up below the national poverty level. Instead, when poverty is acknowledged, bilingual education is identified as one of the culprits. Unz states that bilingual education is a place where children "remain imprisoned" and thus is about "guaranteeing that few would ever gain the proficiency in English they need to get ahead in America." Unz neglects to recognize the fact that even in the cases where English is one's primary language, it does not guarantee economic, political, and integrative success. For example, Native Americans, Native Hawaiians, Chicano/as, and African Americans have been speaking English for generations in this country, and yet the majority of the members of these groups still remain socially, economically, and politically subordinated. Thus, the issue is not simply about language. White supremacy, classism, and other forms of discrimination play a much larger role in limiting one's access to social, economic, institutional, and legal power. Instead of seriously addressing such issues, the English-only coalition serves up myths of meritocracy and life in a melting pot where the patterns of a "common culture" and economic success miraculously emerge if one is willing to submit to their agenda.

What should be particularly disturbing to people is that the national "debates" over bilingual education have very little to do with language (the reason that the word "debate" is questionable here is because virtually all critical voices are either excluded from mainstream national discussions, or forced to find cracks of expression in fringe journals). The general public, that Unz (1999) claims has such a profound grasp of language acquisition theory, seems much more inclined to talk about the people that speak particular languages, rather than the languages that they speak. As witnessed in the controversy over Ebonics, the mainstream discourse has focused on images of African Americans rather than the historical, cultural, and linguistic developments of black English(s). The popular debates thus have more to do with dominant representations of the pros and cons of particular groups, especially blacks and Latino/as. Such a focus not only disregards the multiplicity of other linguistically diverse groups that are at the mercy of powerful anti-bilingual proponents, but it also reveals what is in fact a racialized debate. For example, the English for the Children publicity pamphlet poses the question, "What is 'bilingual education?' " To which it eagerly responds,

> Although "bilingual education" may mean many things in theory, in the overwhelming majority of American schools, "bilingual education" is actually Spanish-almost-only instruction . . .

The word "Spanish" is often strategically used as a code word for the largest, and demographically growing, political force in the country—Latino groups. This racialized marker creates fear among the whites that English-only advocates not only perpetuate but also play on their fears. There is a not-so-subtle play on public fears that the unwashed brown masses from impoverished countries like Mexico and Haiti are on their way to the States. This is particularly dangerous in the hysteria and hatred that has been generated in this country post-9/11.

Unz's own racism can be clearly heard in his comment to the *Los Angeles Times* (1997) when he stated about his Jewish grandparents who were poor and emigrated to California in the 1920s and 1930s: "They came to WORK and become successful . . . not to sit back and be a burden on those who were already here!" (p. 1). Within this racist climate, the appalling conditions faced by both bilingual and English as a second language teachers and their children are by no means conducive to assimilation, let alone selective acculturation, and by no stretch of the imagination, to social transformation.

The Limits of Liberal Advocacy

There are a great many supporters of bilingual education, who have made important contributions in the theory, research, and practice necessary to clearly establish multilingualism as the road to democracy. Many pro-bilingual academics have focused on the importance of understanding the neuropsychological aspects of bilinguality, cognitive models of processing and storing information, assessment, codeswitching, phonemic awareness, and conceptual representations of words. Some supportive researchers have concentrated their energies on the variety of programmatic approaches to bilingual education and the importance of community outreach efforts. Other educators and organizations have set forth research agendas for improving schooling for language-minority children. In addition, such scholars have worked to humanize pedagogical and methodological considerations when teaching linguistically diverse students.

It is certainly important for educators to understand and explore psycholinguistic and sociolinguistic aspects of language acquisition and cultural identity, and to infuse culturally responsive approaches that can accommodate the diversity of students in their classrooms. However, such explorations and responses to the problems of language instruction and learning in the United States also need to name and interrogate the economic, material, and ideological forces and those that create poverty, hunger, discrimination, low self-esteem, and resistance to learning. Well-intentioned studies of reading styles, for example, will not mitigate the antagonistic social relations within which reading takes place in public schools and the society at large.

No Child Left Behind's Monolingual and Phonics Agenda

Since President George W. Bush signed into law the Elementary and Secondary Education Act of 2001, better known as No Child Left Behind (NCLB), high stakes testing has been officially embraced and positioned to be the panacea of academic underachievement in public schools in the United States. The Act engenders a hitherto unheard of transfer of power to federal and state governments, granting them the rights to largely determine the goals and outcomes of these educational institutions. It is ironic, to say the least, that this social movement has emanated from a political party that in the not so distant past called for dismantling the federal Department of Education altogether. As a direct result of this new conservative agenda, school administrators, teachers, communities, and parents are stripped of any substantive decision-making power in the nation's public schools.

Under pressure to produce results on these standardized tests, or face the consequences of cuts in federal resources and funding, school closure, and in some cases law suits, many school administrators have been forced to drastically narrow their curriculum and cut back on anything and everything that is perceived as not contributing to raising test scores. In many cases, this includes the elimination of two-way bilingual education, creative reading, interdisciplinary studies, music and art, community, and athletics programs. Within this "one size fits all" standards approach to schooling, the multifarious voices and needs of culturally diverse, low-income, racially subordinated, and linguistic-minority students are simply ignored or discarded.

Embracing what is in fact an old neoliberal approach dressed up as innovative reform, proponents of this market-driven educational model make use of words and phrases like equity, efficiency, and the enhancement of global competitiveness, to continue to sell its agenda to the public. However, this same political machinery—this synergy between government and the corporate sector—shrouds, in the name of "choice," conservative efforts to privatize public schools and dismantle the nation's teachers unions. Devoted advocates of current legislation also effectively disguise the motivations of a profit-driven testing industry led by publishing power houses like McGraw-Hill, which is the largest producer of standardized tests in the country. In the end, corporate elites of the likes of Harold McGraw III, CEO of McGraw-Hill, who was appointed by President Bush to the Transition Advisory Committee on Trade, will be the only ones to gain from this national obsession with standardized assessment. Speaking at the White House, as part of a group of "education leaders" invited by George W. Bush on his first day in office, McGraw III stated:

> It's a great day for education, because we now have substantial alignment among all the key constituents—the public, the education community, business and political leaders—that results matter.

The results that matter are that corporations like McGraw-Hill gain financially both by selling their materials on a grand scale—a 7 billion dollar take, and by the ways that schools will now guarantee the production of a low and semiskilled labor force that is in high demand in our now postindustrial, service-oriented economy; especially since millions of more lucrative industrial and white collar jobs are being exported by U.S. corporations to nations that pay below a living wage and that ensure that workers have no protection under labor unions and laws that regulate corporate interests and power.

A key characteristic of the new "highly qualified teacher," according to NCLB, is their ability to pass a subject matter test administered by the state. Reducing teacher expertise to a fixed body of content knowledge, middle and high school teachers are expected to meet an extremely narrow range of skill requirements under the new policy. Any concern with pedagogy—not what we learn, but how we learn it—has virtually disappeared. As a direct consequence of this political climate, public schools are being inundated with prepackaged and teacher-proof curricula, standardized tests, and accountability schemes.

Conservatives insist, ad nauseam, that scientifically based research inform and sustain the nation's educational practices, policies, and goals. However, the empirical studies that are used to buttress the Bush agenda, under close scrutiny, are easily stripped of any legitimacy. The well-funded think tanks that produce much of the research and literature to support conservative causes have an obvious, ideologically specific take on these issues, one that is widely supported by mainstream corporate media whose ownership have similar interests.

Perhaps the most strikingly fraudulent use of "scientific research" is the official report signed and circulated by the Congressionally appointed National Reading Panel (NRP), which informs Bush's *Reading First* literacy campaign replete as it is with inconsistencies, methodological flaws, and blatant biases. For starters, Bush's educational advisor when he was the governor of Texas, G. Reid Lyon, headed the NRP. A staunch phonics advocate, Lyon hand selected the panel and made certain that virtually all of the participants shared his views. Curiously, there was only one reading teacher on the NRP. However, by the end of the group's investigation into effective literacy practices, she refused to sign the panel's final report, maintaining that it was a manipulation of data, and that the cohort failed to examine important research that did not corroborate its desired findings.

In the guise of benevolent reform, programs like *Reading First* feed into ultra-conservative hands by limiting federal Title 1 funds to programs and practices that are accepted by the power structure as being grounded in "scientifically based research." This enables conservatives to push forward English-only mandates and a strictly phonics agenda. In other words, if the research that supports multilingual education and whole language instruction has been dismissed by these ideologically stacked panels as "un-scientific," then only nonbilingual and phonics-centered programs will receive federal funding. While there are over 150 studies that clearly support the effectiveness of bilingual education, including the government's own National Research Council Report, and likewise, there is a mountain of work that attests to the limits of a rigid phonics approach to literacy development, this empirical work is rejected by those in power with a mere wave of the hand. In fact, the concept "bilingual education" no longer appears in the Bush education legislation. Students are now referred to as English Language Learners (ELLs).

While the titles No Child Left Behind and English for the Children connote fairness, compassion, and equity, these political campaigns virtually disregard why inequities exist in the first place. If and when fingers are pointed at the causes of poverty and discrimination, these political forces readily blame progressive educational programs and democratic social policies for the country's plethora of problems: academic underachievement, high student "dropout" rates, crime and violence, unemployment, a failing economy, and so forth.

As advocates of the corporate model of schooling hide behind notions of science, objectivity, neutrality, and "universal" knowledge, what is largely missing from national debates and federal and state policies regarding public education is a recognition and analysis of the social and historical conditions within which teachers teach and learners learn; that is, how racism and other oppressive and malignant ideologies that inform actual educational practices and institutional conditions play a much more significant role in students' academic achievement than whether or not they have access to abstract content, a monolingual setting, and constant evaluation.

Rather than addressing these serious issues, conservative educators like Diane Ravitch, Lynne Cheney, and William Bennett—omnipresent spokespersons for the Republican Party—have and continue to argue that attempts to reveal the underlying values, interests, and power relationships that structure educational policies and practices have corrupted the academic environment. Such efforts to depoliticize the public's understanding of social institutions, especially schools, in the name of neutrality are obviously a reactionary ploy to maintain the status quo. It is precisely this lack of inquiry, analysis, and agency that a critical philosophy of learning and teaching should work to reverse.

As the well-publicized, anti-bilingual camp talks little about language acquisition in any edifying depth, it is no wonder that the general population is ill informed. As a teacher told Ron Unz at one of his presentations in the battle over Massachusetts, "You have to win by ignorance!" She also explained that the English-only petition that was required for the referendum to take place was being dishonestly presented to the public by paid solicitors working for English for the Children: outside of a major department store, the representative was saying to passers-by, "Please sign this petition, it will enhance Bilingual Education." Other people have also expressed their disgust after being manipulated by Unz's hired solicitors. There have been endless stories of people outside of stores asking, for example, "Do you want to stop sales tax in the state?"; then stating immediately after, "Also, if you want to enhance bilingual education, please sign here." In California, many people thought that they were simply voting for English when they cast their support for Unz's initiative. In Colorado, the Supreme Court initially invalidated the English for the Children ballot because the language used to explain parents' rights was "misleading." Massachusetts should have done the same thing given the incredibly misleading ballot question that glorified the English-only approach and gave a brief and inaccurate depiction of bilingual education. It should have been legally contested as it claimed that programs that were in fact being radically overhauled by state government would not change at all. The ballot should be honest in what is being suggested—an extremely limited approach to language instruction that does not make use at all of the native tongue, that usurps parents/caregivers' rights to select what is best for their children, and that if teachers are caught using native-language instruction at all to help students through a very difficult process, such educators can be sued and banned from teaching for five years.

Instead of looking to the plethora of scholarship in the area of language acquisition, and encouraging people to do so, Unz's anti-intellectual demeanor invoked the following response to a reporter's question of, "Do young children learn English faster?":

> In fact, it seems to me that if you ask voters that question, I'd guess that probably about 98 percent would say that children learn faster than adults. The only people who would say otherwise are the ones who have read the bilingual textbooks. (Unz 2000)

In order to counter this outright dismissal of any scholarship that supports multilingualism, the debate over bilingual education should not be left in the hands of a savvy politician who is strategically vying for misinformed populous clout through unanswered questions, theoretical ambiguities, and representational manipulations of what is best for children. Educators need to work tirelessly to get substantive information out to the public so that people can in turn protect themselves and their children when ignorance and deceit come, and it will, to your town.

References

Bhabha, H. (1994). *The Location of Culture*. New York: Routledge.
Crawford, J. (2003). A Few Things Ron Unz Would Prefer You Didn't Know About:English Learners in California. <http://ourworld.compuserve.com/homepages/JWCRAWFORD/castats.htm>
McLeod, J. (2000). *Beginning Postcolonialism*. Manchester: Manchester University Press.

Unz, R. (August 31, 1997). As cited in Barabak, M. (1997). GOP Bid to Mend Rift with Latinos Still Strained. *The Los Angeles Times*, Sunday, p. 1.

Unz, R. (November 1999). California and the End of White America. *Commentary*. <http://www.onenation.org/9911/110199.html>

Unz, R. (April/May 2000). The Right Way for Republicans to Handle Ethnicity in Politics. *American Enterprise*. <www.onenation.org/0004/0400.html>

Chapter 14

Live from Hell's Kitchen, NYC

Regina Andrea Bernard

You walk into a classroom full of your new students. You see faces, and multiple bodies, but through quick glances you cannot recognize race or gender, unless there are "tokens" of each, which makes it hard to forget their individual names later during the course of the semester. The difficulty may arise if there are two tokens among waves of the same. Meaning two faces of color among several white faces. During the hordes of your teaching sessions, you begin a bond with the homogenous group except for the two "tokens." During the semester you keep confusing the names of the two tokens. One would think it is easy to confuse the two. They share the same skin color to you, although one is lighter than the other, what appears to you as the same hair texture, they could both be racially "mixed," but the fact is, that while they look alike to you they together look nothing like the rest of the class.

The personalities of the two tokens are so separately distinct. One has fit into your stereotype of the "House Negro" and the other is the "Field Negro." The House Negro responds kindly to your comments and questions, and do not make too much of an academic fuss when objecting. In fact, they often agree with racist statements made about their own group in an attempt to assimilate into the culture of the class's dialog. The Field Negro of the two is a verbal vigilante. She awaits dialog, commentary, and questions like a jaguar awaiting her prey. She takes offense to most of what has been said by the homogenous group, and visually stalks the House Negro. Every time you call her by the House Negro's name, she looks at the House Negro and awaits the answer. When you apologize for calling her by the name of the other student of color, you have isolated her more so than her individual isolation from the homogenous group. The ideas of the two tokens are so contrastingly different, the fact that you could continuously mix them up is no longer a subconscious error, it is a practice of forcing the "other" to remain as such. This act of academic isolation forces the organic intellectual within the Field Negro to come alive and to shine brightly.

The above case example has been an experience of mine during my years as a master's degree student and a doctoral student sitting in classrooms where I am the only or one of two people of color. While some would call the experience "traumatic," it was and is always dangerous and helpful to my own intellectual development process.

What you do not know about school is that these experiences with racial classifications and "subconscious" racist mistakes are an aid in creating a carved-out intellectual space for the "other." Thus, it helps to strengthen the development of the organic intellectual, who brings to the discussion table experience, and social dynamics that are not always negative ones not simply academic references, and statements that remind the lived, and experienced "I *read* that somewhere."

For those who have experienced the "mix up" of names because of race, or color, assimilation is never an option. It is never a reason to want to fit in or become part of the group. It helps to isolate one from mainstream practices of learning and teaching. However, the burden of hard work is continuous and oftentimes depressing. If you are of color, in order to successfully prove yourself as a "traditional intellectual," one who learned all they know through the efforts of being formally schooled, you must go above and beyond the call of duty. I personally have been called to duty since my days as a junior high school student. I still perform my duties to date, as an adjunct, as a student, as a sibling, as two people's children, and as a critical thinker.

Always referencing Hell's Kitchen as the place where I was born and raised back in the 1970s, the traditional intellectuals, have always created a dim picture of what my life "must have" been like. I am tokenized by both race and gender. Traditional intellectuals in thinking of Hell's Kitchen envision rampant drug use and sale, reckless sex, prostitution, loud blacks and Latinos, dirty and unsafe streets, and tons upon tons of welfare recipients. The traditional intellectual envisions these pictures because of what they have read, or what they have heard, or what they have studied. Being a Hell's Kitchen native did show pictures of the above, but only in small vignettes. In fact, my life had never been so good and so educational as the days I spent growing up in the same neighborhood for over 20 years. In the way I have come to make sense of the world, the realities and the fictitious resemblances of what people think they know are the only ways I would only be able to rationalize my own intellectualism by having grown up in Hell's Kitchen, when it was not a trendy place to live, but rather, just my home.

Whether I wanted to become an organic intellectual or not, it was not a choice when combating traditional intellectualism and within its institutions that celebrate it. Unfortunately for the tradition intellectual, the organic intellectual can now find several celebratory instances in that there are clearly several types of intellectualism.

"Organic Intellectualism": the phrase itself can be and has been applied to many individuals possessing its attributes by varying definitions. Coined by Antonio Gramsci, in 1971, his definition was based on organic intellectuals being the rulers of their own organized thought. For Gramsci, organic intellectualism was a refute to the hegemony that shaped most of societal and political thought of his time. While this is a broad definition, many have gone on to further define it, namely Edward Said, W. E. B. DuBois, and others.

One could use many human examples to strengthen the definition of the *organic intellectual*. Malcolm X who gained a particular knowledge through his prison experience is seen as an organic intellectual as opposed to his counterpart Dr Martin Luther King, who received his education traditionally through institutions that "produce" knowledge. Mohandas Karamchand Gandhi traditionally earned his education at the University College of London. There he learned French, Latin, and mastered English. His political "nonviolent" movement toward liberation was birthed through his readings of Henry David Thoreau and John Ruskin.[1] A female counterpart who worked through liberation for women subjected to violence, crimes against humanity, and child-bride arranged marriages, Phoolan Devi, was born and died illiterate. Despite her illiteracy she helped to save many young female children from being forced into arranged marriages, and near her death was inducted into Parliament (Anand, 2002). While all of these people have contributed to the world what was and is still necessary for human liberation and a somewhat balanced humanity, it is rare to find a note on an organic intellectual who served the world through the arts, whether poetry, music, art, writing, or some other form of the humanities.

An Organic Kitchen

In 2004, the "organic intellectuals" perhaps in Hell's Kitchen, as it has come to reshape itself, can be found drinking a half-caf, venti, caramel latte with extra foam in a Starbuck's Coffee Bar, having a draft beer or a *Sex in the City* drink, purchasing their groceries at organic foods shops or farmer's markets, or in front of a "poetry club" smoking clove cigarette after cigarette, beneath stage lights of Broadway with head-attached microphones, signing copies of their book at

Barnes and Noble bookstores, barefoot and afro, or shoed and dreadlocked into their own rehearsed oblivion, that which is trendy.

Looking back toward the 1970s into the mid-1980s, the organic intellectual could be found on his or her fire escape of his or her tenement or apartment building, gathering air conditioning from natural breezes for lack of affording an air conditioner or fan that works. These intellectuals can be found sitting in chairs lined up side by side in front of apartment buildings without stoops, or they can be found collectively appropriating stoops all throughout one block in Hell's Kitchen, El Barrio, Loisaida, and similar neighborhoods. They can be found drinking Pilon or Bustelo or Greek-cupped coffee on those same stoops, where it appears more "ghetto" than the coffee drinker who sips his espresso in an outdoor café. They can be found sitting on the steps of entrances to New York City's subways, writing, reading, while "passing" for homeless. My own grandmother, a Hindu from British Guyana, who died at 95 years of age, could be found telling time but not being able to recognize her own name on paper for being illiterate.

While the construction of experience for the organic intellectual is revered as "ghetto,"[2] many of their trends have been picked up and mainstreamed as part of several new trends. From the wave of sneakers without laces, and sagging/oversized pants without belts (modeled after prison attire) to the continuous reappearance of bell-bottom and flared pants that now cost $95–$118.00),[3] the organic intellectual who could not read patterns but made their own clothing, helped to push trends into mainstream fashion without receiving reparations for their creations of poverty.

Traditional dishes, of curry and rice, rice and beans, platanos dulce or mas duro are still included as part of the "Caribbean Restaurant" cuisine for Zagat's diners, and will run between $15–$25 for a plate of the pauper's dish.[4] While these trends have picked up the pace toward mainstreaming and appropriation, the impoverished organic intellectual is still surprised that their cultural "naturals"[5] are being unsuccessfully, but expensively, mimicked. Starbucks offers the option of various milks and condiments to adorn various coffees. The organic intellectuals boils their canned coffees in a "greca" (Spanish term for coffee pot that sits on the stove rather than in a machine) and is sometimes consumed black or countered with the cheaper condensed milk (sweet milk) or half-half that is half milk and half cream but a whole of saturated fat. Farmer's markets and gourmet or organic food stores are replaced by bodegas, which sell 40 ounces of varying malt liquor, hosiery, and sandwiches all in the same shop, or small grocery stores that sell neighborhood-specific products all of which are highly saturated in fat or just unhealthy. It is not due to a lack of knowledge, it is due to a lack of access. If healthy foods are not stocked in various neighborhoods, it would take a good amount of access in order to achieve a well-balanced diet.

For example, on the strip from 114th Street and 95th Street in Corona, New York (the south side of Queens), there are three supermarkets and several bodegas. At any given time, the products in any of these supermarkets are stale. Meats are redated and repackaged to get them off the shelves, breads are moldy, but the alcohol aisle is always overstocked, and whatever is skim on the shelves, a helpful stock clerk will retrieve your request from the stockroom. At any of these supermarkets, elderly Latino/as await your change in their plastic containers as they overpack your shopping bags. On the same strip between 114th and 95th streets, there are multiple gas stations adjacent to people's homes, apartment buildings, day care centers, and schools. The smell of toxins is unbearable during the summer months. There are also multiple car dealerships all of which house vicious guard dogs that often become loose through the neighborhood because of faulty gates.

Where I grew up in Hell's Kitchen and lived for over 20 years, restaurant were somewhat affordable, and now 20 odd years later, it is geared for tourists who can afford to eat and pay in their choice of currency. Botanicas also termed as "folk pharmacies" (Neal, 2002) and candy stores have been replaced with quaint boutique-like restaurants and coffee shops. The Laundromats have been replaced by patrons of color to white yuppies who "drop off" their wash rather than wash their clothing themselves. A French restaurant offering unpronounceable dishes replaced greasy spoon Chinese restaurants that were my favorite because of its price, and its neighborhood friendliness. Handball courts have turned into dog runs. "Hooker Hang-Outs" are now wineries and lofts for tasting sessions. Everything healthy, and in the summer, everything outdoor.

During the summers in Hell's Kitchen, my childhood friends and I knew to keep away from the ladies who practiced Santeria. Now I wonder where they are? Recently, I took a drive past their apartment building in the Kitchen, and saw markings of the "rosita" still on their walls outside of their buildings, but got wind that they had moved several years before my return. Even Red Apple supermarket has been changed to an upscale Gristedes that is home to what seems like hundreds of young white couples and singles who shop for gourmet products. Growing up, we knew we were in *our* supermarket since the odor of sour milk or a dirty mop that cleaned up the sour milk assaulted us.

During one of my returns to the Kitchen, all the organic intellectuals seemed to have been replaced. Without dismissing the notion that many of them perhaps have passed away (they were the elders of the block when I was a child), it was clear that a wave of traditional intellectualism had replaced them. There were no longer domino tables set up with four old Latino men, with their hats and cabana shirts, slamming pieces down yelling "cohelo alli! (take it there!)" Kids were not running up and down the streets in sweat-filled clothing with sun-matted hair. No one was on my stoop. There were no hungry tons of children of color flooding Saccos Pizzeria for ices and slices. Instead, I was able to immerse myself into a 360-degree view of young white men and women eating in the outdoor option of "Mangia y Bevi." There were no children anywhere. No teenagers anywhere, unless they were working in the supermarkets. No candy stores, no loose-ies[6] for sale. The neighborhoods that were "off limits" to my adolescent and teenage cohort was now flooded with people walking alone or jogging their "Atkins"[7] off.

The parking lot where my father parked his car on a monthly basis was jammed in every corner with BMWs, Porsches, Lexuses, Cadillacs, and Jaguars. I looked for the old maroon Cadillac that my parents owned, and realized even they replaced their Hell's Kitchen mobile with the latest Toyota Camrys. Walking with a friend who is not from the Kitchen, I stopped the car several times and gave him an overview of what I "used to do there" and "what I did here." I asked him if he could imagine about 20 adolescent girls of color singing slow jams on the top of their stoop on a hot August night. The memory connected directly to the organic intellectualism that I speak of here. None of the girls was good in math. In fact, it was the class we struggled with the most. There was no sense of us memorizing the numbers and where they fit in a problem, but we knew each word in every song, and when to use our child-like vibratos, and when to pitch it high or low. We knew how to lace our sneakers straight across instead of crisscrossed, we knew how to read *Right On! magazine*, we knew how much the magazine would cost the group, and how much each person would need to chip in to buy it, but we could not add to save our academic lives.

As mentioned in the previous section, all of us had academic nightmares that we could not beat during the day or at night when it was time to attack the homework. I for one had the toughest time learning how to tell time. My mother bought me a cardboard clock with two giant red hands, and she worked with me everyday and every free moment she had. I still could not get it right. Yet, I was never late. My handwriting was atrocious all throughout elementary school. I never crossed my *t*s dotted my *i*s. I did well in Science, but no one ever knew, because I never raised my hand to answer a question. I was deathly afraid of my science teacher, and my math teacher. They seemed more like drill instructors or correction officers than they did teachers. I detested gym. Every other kid loves gym. My purpose in laying the academic framework here is to accentuate the intellectualism that took place on my block more than it did between the walls of public school, where my teachers adored my "little body" and my "little face" and my "beautiful dresses."

As a backdrop (I was never put up front) member of the National Dance Institute for seven years, I brought my traditional-orchestrated dance steps to my block. After a few minutes, my girlfriends and I had re-choreographed the same sequence, but this time it was in a hip-hop style. My ear was now professionally trained to count the beats in each song in connection to each move I made with my body. The traditional ballet summer camps I attended were nothing like my teenage years spent learning West African dance. I still could not do my math assignments, but I mastered the art of grocery shopping for my mother—always the right amounts of every purchase, followed by the return of exact change. We all (my childhood friends and I) struggled with math and other various academic quests, but we read letters for illiterate elderly neighbors, we served as accurate language translators at various city/state agencies for the parents of friends who did not speak English. We learned how to negotiate adult spaces in our childhoods.

Although our organic intellectualism got us through life in the Kitchen, and helped to create a rich childhood, one that could not be compared with nor replaced by anyone else's suburban or wealthy experiences, it caught a bad rap. With the likes of early teenage pregnancies, juvenile detention halls for some of the boys, drug use and abuse, prostitution, and the era of disappearing, all that we knew organically is what was traditionally studied in our contemporary world of academia. Just as organic intellectualism caught a bad rap because of the end results of not being traditional, organic intellectuals ourselves and/or themselves also received the negative effects of a double consciousness. Academic voyeurs have taken a lifestyle of circumstance and have created a space where organic intellectuals are now and have been for a long time viewed as perhaps trendy to emulate but never attached to any cultural history.

At my Dominican babysitter's (also known best as my *Abuela*, Spanish translation for grandmother) home, I remember dancing with her at the age of 5 to Tito Puente's *Salsa Caliente* and *Ran Kan Kan*. At 5 while I spoke English only at home, and knew Guyanese terms fluidly, I also knew how to salsa and merengue myself all around my abuela's house, especially to *Ran Kan Kan*. She speaks only Spanish, and thus, I spoke only Spanish at her house, and around my Dominican family. I knew the distinctions between Boricuas and Dominicanos, from their language differences, to their dishes to the symbolism of their colored flags, to their native struggles.

In elementary school, the only story that was read to us written by a Latino/a author was *Perez and Martina* (1991) by Pura Belpre. Belpre, the first Latina librarian of the New York Public Library, wrote in *Perez and Martina* about the love life regarding Perez the cockroach and Martina the mouse. It is my only childhood memory of a Latino tale or folklore being read to me as an elementary school student. In junior high school there was none. In high school I suffered through Shakespeare, and in my undergraduate days, there was none that I can really say spoke of the Latino experience that I became aware of in connection to my shared ontology in the Kitchen. I have not mentioned Guyanese or West Indian/Caribbean writers, because school did not give me any of those options to delve into. It was not until I came into the ongoing process of self-awareness that I did seek out my own cultural roots and ties within literatures of the same.

Elementary, junior high, high school, and, even more scarily, colleges are used to determining what students should read and even write about, which restricts what students are allowed to think about. Professors and teachers in traditional-intellectual settings are often the sole creators of what one learns, but organic intellectuals help to decide what is retained. While structure is important, it is oftentimes used as a way to constrict the minds of students in that they are gaining one perspective of a particular topic or subject, and therefore not finding a personal connection to the material. So focused are students on "class aims" and a "do now" enveloped in time management that they are conditioned to pack their bags even before they hear the bell, as teachers remind students of their "assigned" readings for the next class. By the time some students have reached their next class, their minds and concentrations have shifted from one context to the next; how many times does that context place the students themselves at the center and in relation to what they have learned or read? Are these lessons that they can share amongst the girls and boys on their neighborhood block?

As a child I remember seeing a beat up 1960s copy of *Down these Mean Streets* or the Spanish version *Por Esta Calles Bravas* sitting on my bookshelf at home. The book belonged either to my sister or to my brother both of whom were attending high school at the time. As an adolescent and a teenager, I tried to get into the book several times. It was not until my sophomore year in high school that I read the entire book and had to take a deep breath after reading it. Now, it is one of the books that I have been teaching to my students for the past three years, both in the Education department at Hunter College (City University of New York) and the Black and Hispanic Studies department at Baruch College, also a member of the City University of New York.

Down these Mean Streets gave me images of people I had known throughout the Kitchen's roster. In the text, I envisioned people who looked like my neighbors and places that resembled my neighborhood. Although I had a fond affection for *Medea*, I felt she never captured any physical similarities to anyone I had ever known or was bound to come into contact with. Texts like *Medea* were more of an abstract affection that I developed.

During my early years of educational training, I was lucky enough to already have a love of reading prior to the intensity of reading assignments that school demanded of me. I spent a lot of leisure time reading for nonschool assignment purposes. I read everything, but I had a particular affection for Stephen King, R. L. Stine, Judy Blume, Cynthia Voigt, *Archie*, and *Betty & Veronica* comic books, and other readings that did not represent me in any way. Although they did not represent me, these readings were accessible both in the school library and in "young adult" sections of various bookstores. Even if my imagination attempted to place me in the main character's role or setting, it was not possible because of the pictures on the book cover. Each of these books at the time and still now are beautifully illustrated with white teenage girls and boys. During my postschool hours, I oftentimes shopped around bookstores for texts that related more or represented me in closer proximities. Perhaps someone in some novel looked like me, or lived in a neighborhood like I did, or did the things that I, and my friends from my neighborhood, did. By junior high school I had fallen into a relationship with Zora Neale Hurston, Langston Hughes, Wallace Thurman, and the works of Piri Thomas, Nicholasa Mohr, and Nuyorican poetry. The authors reflected something that previous books had not. They reflected images and experiences I saw in myself, and better still, they also reflected people I saw on a regular basis in my neighborhood. Being born and raised in Hell's Kitchen between the 1970s and the 2000s reading became my partner because of the low-excitement levels of activities that were available to young people growing up on my block. From my window in Hell's Kitchen I watched as almost all of my girlfriends became mothers before they were old enough to vote. I watched as populations of adults developed some type of drug or alcohol problem. While I was not on my stoop with my friends playing all types of games, or boy-watching, or smack-talking, I was reading, professionally dancing, learning dispute resolution in after-school programs, and by the time I was 14, I was working at the New York Public Library.

By the time I reached high school, I had read everything that Zora Neale Hurston and Langston Hughes had ever written. Rarely their texts would be taught in school(s), but somehow it always felt as part of a departmental requirement that professors and teachers would assign their texts. Just as a World Literature course would assign the *Bible*, and sections of the *Bhagavad Gita* as part of its global perspective. There was no prefix to these texts, they were just "dumped" into the curriculum, and we as students were to come to our own conclusion. I was utterly disgusted by the likes of *Gulliver's Travels*, and ended up with an "F" in a college Western Literature class because of failing a final exam that was based on Swift's text.

There are several voices and perspectives missing from the world of literature courses as part of various college requirements. In representing my own cultural ontology, I could simply ask of the literature required courses, where is the Guyanese, Nuyorican, Dominican, and other people of color voice? It is no doubt that this project will consume political spaces, and hold several institutions responsible for the continuous silencing of people of color, and thus the continuous formation of a silenced student of color. During my first year of doctoral study, I turned in a final paper in which the professor responded "it seems that you are yelling at people from your paper!" I was shocked. Not at her response, but at her surprise. While people of color, in particular, women of color, have been yelping from their literary pages for decades, why is it that mine is not acceptable? Although I kept my original draft as a trophy of "rage on the page," I was asked to rewrite my paper and authorized to resubmit it for a better grade. One of the conditions regarding a better grade was for me to insert the writings of three white women education specialists. These women represented the white traditional intellectual, and their perspectives on people like me. I had lost a sense of self-representation.

Through my teaching experience at the college level at two City University of New York colleges, and one private institution in Washington Heights, I have used the example of "rage on the page" to formulate a writing style that includes self-representation and expression for my students to feel comfortable in composing. I have developed this method and used outlaw poetry to constantly guide my work, regardless of the focus. The mission is to be blatant in whatever form I choose to compose.

Why does school *not* reflect the personal and learned experiences of youth of color? Thus, why does the traditional form of intellectualism refuse to inculcate the organic intellectualism that students may bring with them? Research has shown that teenagers are more likely to be victims and perpetrators of crimes during the 3 hr postschool day. It is obvious that these youth prefer to look to their norms and social spaces once school is over because of the lack of connection between the hours of school being in session and the hours of when it is over. The traditional intellectual model of teaching and learning asks that students learn to read, write, and work through mathematical problems regardless of their backgrounds and their social situations. While it may seem a fair request of students who do attend school, one has to also think about how students should be able or encouraged to use their organic methods of reading, writing, and working on math problems in connection to what is required of them. Some may say that after-school programs supply the remedy for this connection of organic intellectualism and traditional intellectualism.

However, students have to *see* the potential in what after-school programs can have for them, and essentially have to *want* to be there. Many teens have jobs that require them to begin working directly after school; thus, these students cannot partake in after-school functions. A few hours on the weekend is used as spare time for working youth to engage in their organic and social spaces. From this small amount of time for youth to experience themselves outside of work and school, they are forced to conform and constrict themselves in various mindsets. Homework becomes directly related to school, and school becomes directly related to going to work once school is over. In this framework, youth tend to outgrow their organic intellectualism and social spaces when they are not at school or at work. If school and employment opportunities are not places where youth can offer their organic skills, experience, and ideas, what is left of the young organic intellectual? Young organic intellectuals begins to shed their homemade intellectualism and opt for something a bit more traditional in style. For example, he or she may become embarrassed by an illiterate grandparent, rather than explore the depths of that illiteracy juxtaposed to how the grandparent cannot read, but can tell time, or count money.

Schools have incorporated dialogs about multiculturalism and diversity as a bridge that connects young students of color to curricula that represents them. However, no one asks the students who represent this population the simplest set of questions: "Who are you? What makes you you? What do you know? What did you know before you came to school and what do you know now?" Academics for many New York City youth is built on intimate relationships and friendships within the school, but when it comes to personal connections to what they are learning, oftentimes they must wait until college to experience this. I remember how, when I attended Norman Thomas High School, my love of reading began to grow outside of the school hours because of the impact my English teacher had on me. We read literature fit for Puritans and anyone else that was *not* of color.

I longed to hear or tell my own story. I wanted so desperately to write an essay on what my friends and I did on our apartment building stoop. We knew much about Hell's Kitchen as opposed to the people who now live there, who at that time were fearful of us when we lived there. I wanted to tell of the laundromats, the fire escapes, the break-dancing we competed to master on our block, or our basement of the building. I wanted to tell my English teacher that the drinking scenes in his assigned readings were viewed as social for the characters in the text; however, the reality of my neighborhood would have academicians view those same drinking scenes among people of color as "alcoholism." Wanting to share my experiences, I asked my English teacher, "why can't we maybe do something different but still within the requirements of the reading?" His response, "If you don't like it, you can go play in heavy traffic." I was crushed. My love for reading became resentful of ideas that did not represent any of my experiences. Out of this resentment I began to develop researching skills by seeking texts that spoke directly to my experiences. At one point, my researching resulted in finding nothing, so I began to write my own narrative and organic experiences in poetic forms. I refused to do his assignments after that, and I pretty much refused to attend his class altogether.

Young organic intellectualism, developed in New York City, seems to be so feared by the general public, both visitors of New York and natives. The fear seems derived from the lack of the general public being exposed to this type of knowledge expression. Academics and traditional students are used to reading about these particular places, and may find shame in speaking out if their own neighborhood ends up on a syllabus or reading list. Many do not take the time to understand how it is that young people organize themselves and express what they know and how they have come to know it. Another example from Norman Thomas High School was the issue of "captive lunch." The high school, which is located on Park Avenue in Manhattan, is home to many corporate businesses and corporate Americans. The school was populated with over 90 percent students of color. During lunch, we, for two years, were allowed to eat our lunch outside of the building if we did not wish to eat at the school cafeteria. As New York City police officers watched our every move on the outside of the school during lunch, we still managed to digest whatever we could in our 45-min sprint to the nearest deli or pizzeria. When residents and businesses of the Park Avenue community began to get "nervous" about our presence, which to them appeared in droves, we were forced to eat lunch in the cafeteria from their time of complaint until we graduated.

Imagine thousands of students in one cafeteria at the same time. The absence of our activist and political voices juxtaposed with every other social dynamic (race, age, gender, and our socioeconomic status) lent itself to another birthing of the organic intellectual. Many of the "captive" lunch-eaters worked on rap songs and competed in the cafeteria to a wide audience of listeners and cafeteria employees. Other students learned how to braid each other's hair in various African styles, some wrote poetry and read it to others, and some shared tips on how to beat the system—the parental system through the development of youth-language, also known as "gibberish," which some students spoke more fluently than their requirements of Spanish or French. Mostly all of us were engaging in the steady development of our organic intellectual spaces. The things we could not learn from the classes before or after lunch were being constructed, taught, and memorized in the cafeteria. Those were the lessons we took back to our block for all the rest of the kids to learn.

I look back at my most memorable experiences having to do with learning something. My brother teaching me how to ride a bike, me teaching myself how to roller skate and wanting to become a "roller derby girl," my sister teaching me how to memorize lyrics to songs she would sing to me, me learning how to color-coordinate my 1970s dress code, my mother teaching me how to multiply numbers like eight and nine and tell time on a clock, my father teaching me to cross every t and dot every i and how to be bold to ask teachers questions about lessons without fear. My grandmother teaching me the art of caring for the elderly and the constant process of self-reflection that one must engage in, in order to see one's center, my "abuela" for teaching me Spanish and never having learned English herself. My family for being multicultural and engaging me in the exploration that makes me multicultural as well, so that by the time academia got hip to "multiculturalism" it was already deeply sewn into the fabrics of my life. It was not until graduate school that I again felt the same way about learning things the way I had learned them organically and from my environment and those who shared spaces in that environment as well. Graduate school has taught me the traditionally intellectual terminologies to words and definitions of experiences that I have always had, but never knew that organically acquiring them was an applicable concept in school.

Schools and other traditional institutions of intellectualism have failed youth in their attempt to create a bridge between the two forms of intellectual development. Instead of teaching from tools (experiences) that youth already have, traditional institutions of intellectualism have used what they *think* they know about what youth have. Take the spoken word for example. Spoken word in the Nuyorican tradition has always been about writing and reciting from what you feel and have experienced. Spoken word has now taken on the tradition of mass marketing popularity, such that it is available on HBO, beneath Broadway lights, and it is also being "taught" to many New York City youth by way of after-school organizations. How does one teach not from experience, but about experience? Is that even a concept that one can package and sell?

What happens to the organic intellectual of color that he or she dies so early without ever fully being recognized by the traditionalists? Is it their access to the world or lack thereof? Is it what consumes them during their lives? Is it we, their audience that forces them out of their self-sponsorship? What about their production of knowledge, and our responsibility to help consume what they have produced? While others pick and choose for us as to what we read, write, listen and dance to, and internalize as entertainment, what about the intellectuals whose organic qualities are misused in life and utterly misappropriated in their death? Are they left without any choices but to become traditional intellectuals or die unbeknownst to the world that subconsciously inherits and celebrates their inventions and/or practices?

What you don't know about school is everything you have to find out on your own.

Notes

1. http://www.ccds.charlotte.nc.us/History/India/save/mclanahan/mclanahan.html, Online: August 2, 2004.
2. The term "ghetto" in this section refers to the colloquial definition of those who live "uncivilized" and in "uncivilized places." For example, not having enough money to eat a single meal, but abusing gathered funds to purchase rims for a hatchback automobile.
3. See web sites for stores such as BeBe and Diesel: www.bebe.com and www.diesel.com, retrieved August 22, 2004.
4. See www.citysearch (Cabana Restaurant link), retrieved August 22, 2004.
5. The term "Naturals" is used here to denote everyday life experiences.

6. The term "loosey" is used to denote a single cigarette sold for $0.25 rather than the obligation of purchasing a whole pack of cigarettes.
7. "Atkins" refers to the Atkins diet that consists of low carbohydrates and soaring cholesterol.

References

Anand, V. (2002). *Poolan Devi, a Martyr in the Fight against Caste, Race, and Gender Oppression.* http://www.ambedkar.org/news/phoolandevi.htm

Belpre, P. (1991). *Perez and Martina.* New York: Viking.

Gramsci, A. (1998). In D. Forgacs (ed.), *An Antonio Gramsci Reader.* New York: Schocken Books.

Neal, M. (2002). *Nuyorican Nostalgia.* http://www.popmatters.com/columns/criticalnoire/020423.shtml

Chapter 15

Letting Them Eat Cake: What Else Don't We Know about Schools?

Shirley R. Steinberg

What is being covered up in America's public schools? Perhaps the question should be: what is *not* being covered up? Readers have been given glimpses into conditions, ideologies, policies, and inadequacies in the previous chapters. But what is it that is plainly set in front of us as parents, teachers, and citizens that we are not seeing? Not hearing?

Since the Reagan years, the American public has been indoctrinated into a manufactured notion of safety, family values, back to basics, and scientifically based research. The language of American educational politics has reassured the public that our children are being *taken care of* and that our educational system (the best in the world), is tirelessly engaged in creating good schools. The Orwellian discourse we have been fed is direct, yet meaningless, in determining exactly what it is our schools are doing. *What is happening in our schools* is hidden right in front of us. We see it, we just do not *see* it.

Trotted out en mass, experts, statisticians, and politicos tell us what our children need. Smothered in positivistic and condescending words, bills are passed and more rights are lost. No Child Left Behind never tells us exactly where the children are left behind from and more importantly, where they are going. The bill leaves teachers, parents, and school districts behind—behind piles of bureaucratic measurements that basically blackmail schools into adopting simplistic standards and benchmarks that only ensure that no child gets ahead.

Our school administrators, regional and district offices, and schools of education are complicit in the plainspeech of governmental interference in education. Spinning around with new dictates to measure performance, any social, contextual, or qualitative observations are lost. Humanistic curriculums, ethical school administration, and scholarly teachers are replaced by the need for national recognition, certification, and performance scores. Ironically, the only things not being measured and certified are test creators, textbook publishers, district and regional school offices, administrators of higher education providers, and authors of governmentally sanctioned *scientifically based research*. Just because something is called scientific, it does not mean that it is. Whose science? Whose research?

Sadly, this country (one that proclaims its integrity through a system of checks and balances) has no accountability at the top. Consequently, the top administrators often ignorantly erase what good has been accomplished. And we must ask: *Quis custodiet ipsos custodies?* (Who is guarding the guards?) When those in power intimidate and proclaim mandates, who is there to challenge the origin of these changes?

And when these changes are challenged (which is seldom), how easily is evidence changed and data skewed? When Czarina Catherine the Great of Russia would travel into the countryside and see her subjects, main streets of villages were cleaned, fronts of houses and stores were repainted, and citizens were dressed in new clothes. Her entourage would pass through town and she was assured that her people were living in beautiful towns, well fed, and prosperous. As an educator of graduate students who are teachers in public schools, I am inundated by *new coat of paint* stories. Weekly, my students come to class with tales of mandates from their principals to prepare classrooms for regional or district visits. In classrooms that have not had a budget for paper or books for months, money suddenly appears to replenish tattered books, polish filthy floors, and to fix stopped up toilets. Bulletin boards become the focus as students' work is placed with a grading rubric attached. State and local standards are plastered around the rooms, and students are told to come dressed in their best.

Colleges of education do not fare any better. Recently an urban school of education became victim to the need for national recognition. Determined to acquire a legitimating stamp for teacher education, an entire university system mandated that each school of education had to apply for recognition through the National Council for Accreditation of Teacher Education (NCATE). In a college that has no budget for photocopying, text materials, few full-time faculty lines, and over-cramped classrooms, hundreds of thousands of dollars miraculously appeared to create the illusion that the school of education was, indeed, in compliance with NCATE's standards. There was never a discussion as to whether or not this university already had a *good* school of education, no investigation, no data collected. No one seemed to care what the quality of education for teachers was, only if correct measurements were applied and the right answers given in order to acquire *national recognition*. Witnessed were two solid years of scurrying—administrators, faculty, and students—concerned with one thing: *National Accreditation*. Discussions of good teaching, smart curriculums, students with needs, good scholarship, and research were halted. Meetings were replaced with forms to fill out, benchmarks to make, rubrics to create, syllabi to alter to comply with NCATE's standards. The competent faculty of this school became witness to the dissolution of competent departments, the exhaustion of fine professors, and the mystified queries of students who were told that the 17-page syllabus was not *really the course syllabus, just something to comply with the NCATE standards*. The time and money spent on *national recognition* did not go unnoticed by the faculty, but it did go unchallenged. And so it goes in many schools of education.

Ironically, schools of education (Madison, Wisconsin; NYU; UCLA, and Berkley, among others) that consistently rank the best in the nation, are not NCATE accredited. They have clearly stated that they believe the time and money spent on a certification by a for-profit organization is not what good teacher education is about. Or is it ironic? Perhaps the pressure from NCATE contributes to the inability to rank highly due to the time and money spent on bureaucratic marking and accounting sheets.

What exactly is "national accreditation?" What is it we do not know about it? The web site fills us in:

> Teaching children—to recognize letters, to read for the first time, to understand how a tree grows—is one of the most important jobs in America. The nation's future depends, in large part, on how well it is done.

NCATE is the profession's mechanism to help establish high-quality teacher preparation. Through the process of professional accreditation of schools, colleges, and departments of education, NCATE works to make a difference in the quality of teaching and teacher preparation today, tomorrow, and for the next century. NCATE's performance-based system of accreditation fosters competent classroom teachers and other educators who work to improve the education of all P-12 students. NCATE believes every student deserves a caring, competent, and highly qualified teacher.

NCATE currently accredits 588 colleges of education with over 100 more seeking NCATE accreditation.

NCATE is a coalition of 33 member organizations of teachers, teacher educators, content specialists, and local and state policy makers. All are committed to quality teaching, and together, the coalition represents over 3 million individuals.

The U.S. Department of Education and the Council for Higher Education Accreditation recognize NCATE as a professional accrediting body for teacher preparation (http://www.ncate.org/public/aboutNCATE.asp).

Naturally we are not told that it is the *only* game in town. Other organizations have made a run at competition, but this multimillion dollar for-profit group has intimidated the teacher education profession into compliance. This is not a regulatory agency, one that answers to the parents, students, and teachers (the obvious constituents), rather, it is an agency that charges schools hundreds of thousands of dollars to become *nationally recognized*. By whom? By them. (Each state has its own teacher certification and accreditation, NCATE recognition is not required.)

"Fixing" teacher education in the United States consists of "expert" agencies coming in to evaluate programs and then applying their stamp of approval or disapproval upon the school. When the expert agencies or regional/divisional board members come to visit our public schools, the walls have been painted, the students and teachers prepped, and all appears well.

In the context of writing this chapter, I have been informally speaking to teachers and asking them what they feel the public does not know about schools. Here is a summary of comments I heard:

- I was hired to work in an Academic Intervention Program as a push-in teacher for reading. I was used and abused as a substitute teacher. The day the Regional Supervisor visited the school it was announced that I was the Academic Intervention Team Leader! I was mortified. The program was a façade and I hated being part of it. The next day I was back to my substitute position. The injustice of being held accountable for things of which you had no control.
- Administrators want their students' work and classrooms to be "pretty." They're only concerned with how the room and walls look. There is no concern for the validity of a child's education or the process in creating his or her work.
- Schedules aren't altered to suit the needs of teachers in regard to classroom management. There are no smart transitions between subjects.
- Our region pays a fortune for a consultant program {Australian educators brought to New York to mentor teachers]. They don't receive the training we do, they are not even certified to teach our students and cannot be left alone in the classroom.
- Parents are not informed that they can demand that their child can be tested for special services, and once the testing is done, the students must be taught to meet their needs. When they are tested, the tests and placements can drag on for months.

WHAT ELSE DON'T WE KNOW ABOUT SCHOOLS?

- Children are not as safe as the schools want the parents to think. There are violent children being allowed to remain in classrooms even after they have hurt another student or even a teacher. The region is too afraid of lawsuits to remove seriously troubled and dangerous kids from the classroom. Naturally, all children have rights, but it becomes difficult to teach when the teacher is afraid of a student, and the student has free reign to cause injuries.
- In our school, the reading specialist was laid off as the school determined it needed to invest in new bookcases and furniture for the school. Our low-achieving school now has no reading specialist, but lots of new bookcases!
- I work with autistic children. My students have the right to have a one-on-one paraprofessional in order to stay in my classroom. However, if the parents don't request one, or sign off on not needing one, the school saves money. My principal encouraged a parent to disallow a para to help their autistic child. The child manifested very difficult behavioral problems, I called the parents and they told me they had signed off on the para because they were not told it was their right to keep one with their child. When I brought it up to the administration, I was disciplined as I had told the mother about her son's rights.
- What don't people know about schools? Do you want a list?
- No gym or recess (even though gym is state-mandated)
- None of the computers work—and if they get fixed, we still wait to get an Internet hook-up—I mean wait—like three years
- We have no city or state-leveled libraries (they are city-mandated)
- We are forced to use Columbia Teachers' College-generated curriculum, indeed our school pays dearly for them, *but*, we have no resources to purchase the materials needed to use this curriculum
- When a teacher is sick or absent, we cannot ask for a substitute, the class is split-up and kids are placed in other classes.
- Teachers are expected to conduct workshops with no preparation, no notice, and no resources
- We have rugs in our classrooms, dirty, torn, hazardous rugs
- We are told we can't use the chalkboards, yet no dry erase boards have ever been installed, they are "coming."
- Our dirty little secret is that in grades 3–5 the children are tracked. The principal sorts the classes based on their standardized exams. Those children who scored well are placed in the "top" class. These children are labeled "gifted." Some children are placed in the "top" class because they are well-behaved. The favored teachers get these classes. New or unpopular teachers get the "low" classes.
- Our school paid $130,000.00 for a playground, and $10,000.00 for a mural on the wall. But we have no paper in our school for the year, no erasers, nor markers for the dry erase boards.
- In our school supplies are hoarded and not given out to the students (pencils, paper, twine, soap . . .).

- Students do not have any extracurricular activities.
- Teachers are expected to do work before and after class in order to function during the day. Taking a lunch period is frowned upon.
- I have no books for my classroom library. But my principal has brought in two staff developers. We also have new painted walls in the hallway and laminated charts for the halls.
- We cannot duplicate anything. One teacher was given a used copy machine to keep in her room and use, but no one else was supposed to know. We are not allowed to make overheads as there are no materials.
- We have clean floors, bathrooms, and classrooms—*only* when visitors come.
- Some teachers are being paid per session to take attendance for test-prep classes after school. They do not have a class, they are paid a salary just to gather attendance sheets.
- I teach in a SUR school (school under review), we are considered one of the worst schools in the city. Our _____ team just came in first place in all of New York City, out of over 75 other schools. They are eligible to go to the world championship this year. These kids have never won anything, gone anywhere, never been told they could be successful. Now they won this amazing competition and there is no money to send them to the worlds. However we pay a fortune for Columbia Teachers' College to come in twice a week to tell us how to teach. The kids are scrambling around trying to raise money—do you know how hard it is to make money when the kids are from such a poor school? Who can donate? Who can buy baked goods? They probably won't go.
- I have been told to hang at least twelve efficacy posters in my classroom, these promote happiness, hard work, and the goals of the school. I teach first grade and the kids can't read these posters. Yet four people have come to visit my room to make sure they were posted.
- I have been given a digital camera and software, but I have no working computer. I got lots of handouts on paper about the camera and the software—yet we have no paper to use for our classwork.
- I teach in a class for emotionally-disturbed students. It is a small class and I manage it with a paraprofessional. Recently two non-English speaking students and one learning-disabled student were placed in the class as there was no other place to put them. Within days they were manifesting behavioral problems—I believe it was because they were placed in a class with children with serious mental problems, some of whom are violent—they shouldn't be in my class.
- I have always taught third grade. I prepared my classroom all summer for my new class. On the first day of school I was told that I wouldn't have third grade. I removed my things from my classroom, was moved to sixth grade. I have never taught sixth. After three weeks, I was moved again, and then again. In the winter I announced that I

- was pregnant, I was removed from my classroom and assigned to "float" and cover classes.
- I have taught English for ten years in this school. I have had the same textbooks that entire time. I attended this school myself and we used these texts in my English class seventeen years ago. The books talk about the fact that "someday, computers will be used even in the home"; and that "people will be able to carry portable phones without wires with them." My students think the books are a joke. And I am supposed to prepare them for graduation and college?
- Our school has paid a fortune for Australian consultants. The problem is that we are in the hood, no joke—our students are from the projects, they are from different cultures, races, and the one thing they have in common is that they are truly urban. I was assigned a consultant for reading, she came in, denounced all my library books as inappropriate, told me that my needs for multiculturalism were incorrect—she told me she was from a rural area of Australia, and had never been to a city until she came to New York. Is this a joke?

Unfortunately, none of this is a joke. Blame is squarely fixed on teachers in the United States, and administrators and school boards remain unaccountable and paranoid. No one guards the guards. In 1983, the Commission on Excellence was formed by the Department of Education. Members of the commission were charged with determining the state of America's public schools. A document, *A Nation at Risk*, was created as a result of this inquiry. Bottom line, the commission led by Republican William Bennett, announced that the problem with America's schools is . . . America's teachers. It becomes the ultimate example of *blaming the victim*. The women and men who work over 6 hr a day in difficult conditions were the bad guys, not the administrators or politicians who controlled the teachers. This report allowed an organization such as the NCATE to become the big brother to schools of education, and with a top-down assessment style, teacher certification has continued to emulate the obsession with measurement, numerical accountability, and a skewed version of *scientifically based research*. As one can see by the summary of statements by public school teachers, even the basic needs of a classroom are not being met. How can one begin to measure students accurately, if learning conditions are not even minimal? Themes of abuse and the purposeful ignorance of administrators emerge through my notes from the comments of teachers. Money and funding become a groundwire for many of the illnesses that pervade our schools. Where is the money? Teachers speak of mismanaged funds, and funds spent through cronyism—money that they never see.

Make no mistake, no one in this book is advocating a lack of standards. Rather, we are arguing for stronger standards. Instead of basing our public schools' needs on positivistic and disconnected observations and measurements, we must demand to have accountability on every level. We must start with search for political advocacy that recognizes the complex and diverse needs of students and teachers. We must join grassroots efforts to keep administrators accountable. And most importantly, we must make sure that parents and teachers, alike, understand their rights and the rights of students.

It is amazing how a small voice of inquiry can reach to high levels. I recall our daughter Bronwyn coming home with a sheet of paper that gave instructions to parents on how to prepare our children for standardized tests. The paper instructed us to:

> Make sure children went to bed early to be well-rested
> Make sure children were given a full breakfast
> Make sure children were at school on time

It was not lost on either Bronwyn or her parents that the school had never made these admonitions before. She noted that: "The school only cares that I am well-fed and rested when they have achievement tests to give. The higher the score, the better our school looks." We sent a note to the school requesting that our daughter be exempted from the exams. The school kicked up a fuss; however, upon investigating our parental rights, they found that we, indeed, had the right to disallow her from taking exams, *and* that the school must provide alternative pedagogical supervision. She was able to attend school and occupied herself doing projects in the library. We immediately saw a change in her affect and lack of nervousness that traditionally accompanied her when she had to take standardized exams. We informed other parents about what we learned, some followed suit, but many others were *scared* (their words) to exempt their children. Perhaps it is time that parents recognize that they do have a say in their sons' and daughters' education. And perhaps it is time that teachers recognized that they also have rights. Teachers' unions do exist, but very little investigation has been done into the ties between union elite and those that receive contracts. Again and again, the teachers I spoke to mentioned that their union rep was absolutely not interested in their issues. Just how deep do the wells of nepotism, manufactured expenses, and hidden resources go?

Stringent penalties should be enacted when an administrator or region denies teachers the right to speak to parents and the public about schools. Districts and regions should be encouraged to visit schools without prior announcement. Schools should be observed as they are, not as they are painted for a particular visit. We must become informed and observant participants in the education of our children and youth.

What is it that we do not know about schools? A hell of a lot. *Quis custodiet ipsos custodies?* It is time that we begin to guard our guards.

Index

1984 Cable Act, 159
9/11 (September 11, 2001), 125, 143, 253

ABC, 160
ABCTE (American Board for Certification of Teacher Education), 6
A Beautiful Mind, 186
Accountability, 3, 79
Achievement tests, 96
ACORN, 123
ACT, 95–96
Addison, Joseph, 185, 187
Adorno, Theodor, 131–132
AES (American Eugenics Society), 172
Aesthetics, 15
Akiko, Yosano, 199
Alliance for Academic Freedom, 125
Alliance for the Separation of School and State, 2
Alta Vista, 137–138
American Coal Foundation, 132
American dream, 110
American Enterprise Institute, 11
American Journal of Public Health, 146
American Library Association, 160
American Press Institute, 160
American psyche, 104
Analytical techniques, 76
A Nation at Risk, 94, 127, 283
Anthropologie, 164
Anti-Semitism, 201
Archie, 269
Assessment, 99
Assimilation, 262
Assimilative policies, 249
Atwater, Lee, 2
Austen, Jane, 199
Authentic assessment, 81, 83–85
Aviles, Arthur, 237
Ayers, William, 82

Baby-Sitter's Club, 148
Balzac, Henri, 183
Banking education, 71
Barbara Bush Foundation for Family Literacy, 18
Barnes and Noble, 263
Barrett, Elizabeth, 200
Battersby, Christine, 200
Behavioral psychology, 106, 109
Behaviorism, 105, 107–108, 110
Belpre, Pura, 268
Bennett, William, 257, 283
Betty and Veronica, 269
Bhabha, Homi, 251
Bhagavad Gita, 270
Bilingual education, 258
Binet, Alfred, 171–172
Blume, Judy, 269
BMW, 266
Bourdieu, Pierre, 175
Bracero Program, 250
Brave New World, 92
Briant, Kobe, 129
Bridgestone-Firestone, 136
Bridwell, Norman, 147
Brigham, Carl, 171
British Broadcasting Service, 168
British Infant Schools, 215
Brown v. Board of Education, 173
Burger King, 133
Bush administration, 12, 16, 20
Bush, George W., 4–5, 16–19, 112, 118, 137, 254–256
Bush, Prescott, 18
Bushian educational science, 20, 22–23
Business Week, 144

Cable in the Classroom, 159
Cadillac, 266
California Achievement Test, 96

INDEX

Canadian Broadcasting Company, 168
Capitalism, 109–110, 127–128
Captain Underpants, 154
Caryle, Thomas, 187, 195, 200
Catherine the Great, 278
CEO, 18, 122, 163, 255
Channel One, 145, 155–159, 161–163
Character education, 134
Charles Schwab, 154
Chauncey, Henry, 178
Cheney, Dick, 11
Cheney, Lynne, 11, 257
Children Now, 160
Children's Defense Fund, 122
Children's Television Act of 1990, 148
Chomsky, Noam, 242
Cisneros, Sandra, 223–224, 226–227, 236–237
Citizenship, 167
Civic skills, 153
Civil Rights Movement, 16
Civil Service, 177
Civil Service Reform (England), 169
Civil War, 97
Class-biased ideology, 7
Class politics/issues, 7–8
Classroom evaluation, 90
Clifford The Big Red Dog, 147
Clinton, Bill, 148
CNN, 158, 160
CNN Presents Classroom Edition, 162
CNN Student News, 158, 160, 162
Cockburn, Cynthia, 228
Cognition discourses, 178
Cognitive complexity, 77
Cognitive and psychometric psychology, 185
Coke, 133
Cold War, 172
Collateral learning theory, 235
Colorado Supreme Court, 258
Columbia Tri-Star Home Video, 148
Comedy Central, 145
Commercialism, 157
Commission on Excellence, 283
Committee of Concerned Journalists, 142
Committee of Ten, 172
Commonality, 248–249
Communications systems, 8
Communism, 4
Competency, 81
Complexity, 70

Complexity of knowledge, 76
Conant, James Bryant, 178
Concentration of energy, 201
Congress, 16–18, 92, 97, 117
Conservatives, 172
Constructing the Political Spectacle, 117
Constructivist learning theory, 106
Consumption, 139
Content-based knowledge, 89
Content model, 95
Content standards, 95
Cops, 129
Core knowledge, 90–92
Corporate influence in education, 120
Corporations, 128, 136–138
Council for Higher Education Accreditation, 280
Counts, George, 131
Crawford, James, 247
Creative diversity, 74
Critical knowledge, 24
Critical literacy, 216
Critical media literacy, 138
Critical pedagogy, 180
Critical theory, 3, 179
Critical thinking, 77
Critical thought, 120
Cronyism, 19
Cultural anthropology, 201
Cultural capital, 175
Cultural literacy, 94, 98
Cultural Literacy: What Every American Needs to Know, 73, 94
Cummins, Jim, 247
Current Events, 145
Curricular standardization, 12
Curriculum, 89, 132, 138
Curriculum alignment, 83
Cyberjournalist.net, 160

Darwin, Charles, 193, 200
D'Souza, Dinesh, 251
Darwinian Theory, 187
Data, 23
De Stael, Germaine, 200
Democracy, 9, 27, 86, 105, 109, 113, 131–132, 138–139, 141, 146, 152, 166–167, 173, 196, 204, 248
Democratic education, 173
Democratic society, 152
Department of Education, 17
Deprofessionalization, 6

Descriptive processes, 15
Deskilling, 73–74
Devi, Phoolan, 263
Dewey, John, 10, 109, 131–132, 174
Diderot, Denis, 185–186, 203
Dieckmann, Herbert, 185
Disney, 150, 156
DISTAR program, 19
Diverse teaching strategies, 77
Diversity, 211, 213–216, 220, 271
Doherty, Christopher, 19
Dominant culture, 223, 236
Dominant ideological activity, 3
Down These Mean Streets, 268, 269
DreamWorks Interactive, 148–149
DuBois, W.E.B., 263

Ebonics, 253
Economic capital, 176
Edelman, Murray, 117
Educating for Diversity, 213
Educational data, 21
Educational evaluation, 81
Educational researchers, 14
Educational science, 15, 23, 25
Educational Sciences Reform Act, 16
Education Press, 145
Einstein, Albert, 183
Elementary and Secondary Education Act of, 2001, 254
Elliot, Charles, 172
Emerson, Ralph Waldo, 187, 194–199, 203–204
Empirical knowledge, 23
Engles, Frederick, 198
English for the Children, 241, 243, 246, 258
English Language Learners, 256
Enron, 18
Enterprise Institute, 156
Entertainment media, 145
Epistemology, 26
Equity, 4
Escalante, Jaime, 251
ESPN, 129
Essentialist movement, 91
Essential knowledge, 90
Ethnocentrism, 11
ETS (Educational Testing Service), 120
Eugenics, 178
Evaluation(s), 28–29, 70, 78, 80, 85, 89, 100

Evaluation experts, 89
Evaluation plans, 79
Evidence-based research, 26
Evolutionary theory, 201
Experience, 15, 92
Experiential knowledge, 25
Extreme Makeover, 129
Exxon, 131–132
ExxonMobil, 132
Eyewitness Kids News, 147, 161–162

FACTS.com, 147
Failure to Hold, 139
Fast food restaurants, 134
FCC, 148
Fielding, Henry, 187, 194–195
Finding Forrester, 186
Finn, Chester, 27, 94, 97
Firestone tires, 136
First Gulf War, 117
Ford, 136, 160
Ford Explorers, 136
Fordham Foundation, 27
Formative assessment, 78–79
Forms of Capital, 175
Fortune 500 CEOs, 18
Fowler, Mark, 146
Fox Kids Network, 148–149
Fox News, 5, 129
Frank, Anne, 199
Frankfurt School, 131
Free market ideology, 109
Free People, 164
Freire, Paulo, 71, 104–105, 131–132, 180
French Enlightenment, 187
French Revolution, 197
Freudian Theory, 187
Frieden, Kenneth, 184
Frontline, 162, 168
FTAA, 142
Fundamentalist Christians, 112
Fundamental knowledge, 27

Galton, Francis, 186–187, 192, 203
GAO (U.S. General Accounting Office), 5
Gardner, Howard, 98
GATT, 142
Gay Rights Movement, 16
General Electric, 159, 161

INDEX

General Motors, 154
Genius, 186
Geoffrey's Reading Railroad, 150
Ghandi, Mohandas Karamchand, 263
Ghettopoly, 164
Global economy, 120, 142
GMAT (Graduate Management Admissions Test), 171
Goethe, 195–196
Golden Marble Awards, 150
Goodman, Yetta, 215
Good Will Hunting, 186
Google, 159
Goosebumps, 148, 150
Gordon, Winsome, 221
GPA (Grade Point Average), 111
Grade inflation, 81
Gramsci, Antonio, 263
Grant, Carl, 213–214
GRE (Graduate Record Examination), 171, 190
Great American Book Fairs, 148
Greene, Maxine, 224, 226, 233
Gresson, Aaron, 16
Gulf Oil, 132
Gulliver's Travels, 270

Habitus, 175
Hamann, Johann Georg, 200
Haraway, Donna, 132
Harris, William, 131
Harry Potter, 154
Hauser, Robert, 177
Hayne, Richard, 163
HBO, 148, 274
Heidegger, Martin, 129
Heritage Foundation, 6, 10, 156
Herrnstein, Richard, 93, 177
High culture, 236
Higher order thinking, 76
High poverty schools, 10
Hildegaard of Bingen, 199
Hirsch, E. D., 73, 94, 98, 100
Historical context, 15
Hoch, Hannah, 202
Homer, 185
Horkheimer, Max, 131
House Resolution 3077, 27
Hughes, Langston, 269–270
Humanity, 109
Hundt, Reed, 148

Hurston, Zora Neale, 269–270
Hutchins Commission, 167
Huxley, Aldous, 92

IBM, 154, 160
Ideology, 3
I Have a Dream, 150
Indigenous knowledge, 73, 90
Individualism, 169
Industrial Areas Foundation, 123
Information Age, 110
Ingenius, 152
Inherent intelligence, 184
Institutionalized racism, 175
Intelligence, 178, 201
Interfaith groups, 123
Interpretive analysis, 15
Iowa Test of Basic Skills, 96
IQ scores, 12, 93, 111, 171–172, 177–178, 190, 203, 231
Iraq invasion, 125
ISLLC (Interstate School Leaders Licensure Consortium), 120

Jaguar, 266
James, William, 92
Jefferson, Thomas, 103
Jegede, Olugbemiro, 235
Jeopardy!, 100
John F. Kennedy Center for the Performing Arts, 148
Johnston, David Cay, 134
Joyce, James, 183
JROTC (Junior Reserve Officers Training Corps), 124–125

Kaiser Family Foundation, 165
Keller, Helen, 199
Kidsnewsroom.org, 159
Kids Power Marketing, 130
KidsSmart Educational Technology, 154
King, Martin Luther, 122, 263
Kingsolver, Barbara, 105
King, Stephen, 269
Kleenex, 154
Knowledge, 21, 71, 73–75, 84, 92, 96
Kraft Food, 135
Kraton, 146
KumaWar: The War on Terror, 167

LAD (Language Acquisition Device), 242
Langley High, 2
Lau v. Nichols, 248
Lawrence, Jacob, 237
Lay, Ken, 18
Leadership, 116
Left-leaning politicians, 2
Letterman, David, 129
Lexus, 266
Liberal ideology, 91
Licensing Industry Merchandizing Association, 149
Lifetime Learning, 146
Lifeworld, 115
Literacy, 18
Literacy skills, 153
Lorber, Michael, 98
LSAT (Law School Admissions Test), 171
Lucky Book Club, 147
Lynch, Jessica, 118
Lyon, G. Reid, 256

Madam Curie, 199
Madison, James, 166
Madlenka, 218
Malcolm X, 263
Manufactured crisis, 117
Marketers, 130
Market Place, 135
Marketplace power, 152
Marxist critique, 187
Marx, Karl, 198
Mass communication corporations, 8
Mattel Inc., 9
Mayan codices, 90
MCAS (Massachusetts Comprehensive Assessment System), 244
MCAT (Medical College Admissions Test), 171
McDonald's, 119, 133, 150, 156
McDonald's Corporation, 149
McGraw, III, Harold, 18
McGraw-Hill, 17–19, 255
McGraw, Jr., Harold, 18
McLaren, Peter, 128, 131
McLeod, John, 252
Medea, 269
Media, 138, 143
Media literacy, 8
Medicaid, 5
Medicare, 5

Merelman, Richard, 166
Meritocracy, 177, 179
Microsoft, 148, 160
Mill, John Stuart, 193
Minority Report, 130
Miramax, 148
Mode of consciousness, 200
Mohr, Nicholasa, 269
Monopoly, 164
Montaigne, 196
Moore, Michael, 164
Morrison, Toni, 199, 236
Mountain Dew, 133
Moyers, Bill, 168
MSNBC, 159–161
MSNBC.COM, 149
MTV, 163
Multiculturalism, 271
Multilingualism, 248, 258
Multiple contexts, 77
Murray, Charles, 93, 177

NAACP, 150
Nader, Ralph, 163
NAEP (National Assessment of Educational Progress), 97
NAFTA, 142
Napoleon, 196–197
National accreditation, 279
National Council on Educational Standards and Testing, 97
National Educational Goals Panel, 97
National Football League, 149
National Geographic Society, 132
National Governor's Association, 97
National History Standards, 11
National Public Radio, 135
National Reading Panel, 256
National Research Council, 21, 256
National Teacher of the Year Award, 149
Native Son, 233
Nazis, 202
NBA, 129
NBC, 148
NCATE (National Council for Accreditation of Teacher Education), 278–280
NCLB (No Child Left Behind), 1, 4, 13, 15–16, 18, 20, 89, 94, 96, 105, 115, 118, 125, 166, 253, 257, 277
NCLR (National Council of La Raza), 245

Nelson B. Heller and Associates, 150
Nepotism, 169
New Line Cinema, 148
New Video, 148
Newspaper Association of America, 151
Newsweek, 147, 161
Newton, Issac, 183
New York State curriculum, 95
Nick News, 162–163
Nickelodeon, 148
Nickelodeon Game Lab, 166
Nick News Special Editions, 163
NIE (Newspapers in Education), 151–152
Nike, 129, 154
Nixon administration, 161
Nixon, Richard, 97
No Excuses: Lessons from 21 High-Performing, High-Poverty Schools, 6
Normative assumptions, 26
Normative knowledge, 24
NCR, 21
NRP (National Reading Panel), 18–19
Nuyorican poetry, 269, 274

Objectives, Methods, and Evaluation for Secondary Teaching, 98
Objective testing, 85, 90
Objectivity, 14
Oliver, Mary, 222
Olsen, Tillie, 236
Ontological knowledge, 24
Open Court Reading Program, 19, 116
Oprah's Book Club, 148
Organic intellectual, 262–263, 266–267, 271, 273–274
Ormerod, Susan, 228
Outsourcing, 142

Paige, Rod, 18, 118
Paramount Pictures, 148
Parliament, 263
Pascal, Blaise, 186–193, 204
Patriotism, 167
Pavlov, Ivan, 104
PBS, 160–161, 168
Pepsi, 133
Perez and Martina, 268
Perfectly Legal, 134
Performance model, 95
Performance standards, 95
Performance tests, 84

Personality, 201
Phonics, 256
Piekhanov, George, 194
Pinker, Steven, 242
Pioneer House, 162
Pizza Hut, 133
Plato, 196
Plutarch, 195
Pokemon, 138
Polanyi, Michael, 176
Polaroid, 154
Political spectacle, 117
Politics of information, 9
Popular culture, 138, 167
Por Esta Calles Bravas, 268
Porsche, 266
Portfolios, 98
Positivism, 11–12, 15, 17, 19, 25
Positivistic aptitude test, 13
Positivistic evidence-based research, 12–15, 21
Positivistic methods, 22
Positivist research tradition, 11
Poverty, 257
Power, 3, 73, 152
Predatory culture, 128–130
Primate Visions, 132
Primedia, 146, 161
Private schools, 1, 5
Privatization of education, 2–3, 6, 8
Privilege and power, 12
Prodigal Summer, 105
Production of knowledge, 11
Professionalism, 115
Professional practice, 120
For-profit corporate schools, 9
Progressive classroom, 106
Progressive education, 174–175
Progressive ideology, 91
Progressive/liberal ideology, 99
Project Head Start, 214, 220
Proof, 186
Proposition, 187, 209, 227, 241, 247, 251
Public Broadcasting Act, 161
Public Broadcasting Corporation, 162
Public Broadcasting System, 148
Puente, Tito, 267
Punkvoter. com, 163
Purpose of schooling, 28

INDEX

Qualitative insight, 15
Quantitative methods, 14
Queen Christina of Sweden, 187, 189
QVC, 154

Race, 262
Race and genius, 201
Racine, 185–186
Rancho Rio High school, 125
Rand Corporation, 160
Raphael, 198
Ravitch, Diane, 94, 97, 257
Reading First, 16, 22, 256
Reagan administration, 127, 161
Reagan, Ronald, 248, 277
Reflective-Synthetic knowledge, 25
Regents Competency Test, 96
Regents Examination, 96
Reggie Award, 150
Reliability, 98
Religion, 2
Research methods, 13, 16
Resnick, Lauren, 76, 95
Reynolds, R.J., 135–136
Right On!, 266
Right-wing advocacy of basic skills, 10
Right-wing agenda, 16
Right-wing education, 3–4, 9
Right-wing educational politics, 27
Right-wing groups, 1
Right-wing ideological success/goals, 3–4, 7
Right-wing operatives, 2, 22, 26
Right-wing policies, 26
Right-wing politics of knowledge, 2
Right-wing think tanks, 1
Rigor, 3
Rigorous standards, 90
Ripple, 154
Ripplewood Holdings, 146
RJR Nabisco, 146
Rodriguez, Richard, 251
Romanticism, 186
Romer, Roy, 97
Rove, Karl, 2
Ruskin, John, 263

Said, Edward, 263
Santeria, 265
SAT (Scholastic Aptitude Test), 93, 95–96, 98–99, 111, 171, 190

Scapp, Ron, 128
Schlafly, Phyllis, 163
Schlosser, Eric, 130–131
Schön, Donald, 25
Scholar practitioners, 15
Scholastic.com, 161
Scholastic, Inc., 149, 151, 153–154, 156, 165
Scholastic Magazine, 145
Scholastic Productions, 148
Scholastic Publishing Company, 147
Scholastic school group, 149
School choice, 2
School testing, 98
School vouchers, 2
Schumpater, Joseph, 127
Science wars, 16
Scientific educational knowledge, 20
Scientific knowledge, 12
Scientific management, 92
Scientifically based research, 256, 278, 283
Scientific testing, 92–93
Searching for Bobby Fischer, 186
Sex In The City, 263
Shakespeare, William, 185, 196
Shaw, Bernard, 191
Shelley, Mary, 200
Shell Oil, 146
Sis, Peter, 218
Situated learning, 100
Situated thinking, 100
Six Flags Theme parks, 154
Smith, Adam, 186–187, 190–193, 195
Social class hierarchy, 81
Social Darwinists, 171, 178
Social efficiency, 173
Social Security, 5
Social theory, 193
Social theory of authorship, 188
Socioeconomic status (SES), 176–177
Spanish Jesuits, 90
Special Edition, 162
Spoken word, 174
Sputnik, 94, 172
Standardization, 116
Standardized curricula of right-wing schools, 10
Standardization of curriculum, 90
Standardized evaluation, 83
Standardized high-stakes testing, 80, 110, 111, 116

INDEX

Standardized testing, 10, 14, 21, 71, 74, 80, 82, 89, 172, 255
Standardized thinking, 100
Standards, 3, 28–29, 70–71, 77, 85, 120
Standards and accountability models, 79
Standards of complexity, 74, 76–78, 83–84, 86
Standards movement, 116
Standard test, 89
Stanley, Thomas J., 170
Starbuck's, 263
Stein, Gertrude, 199
Sternberg, Robert, 99
Stine, R.L., 269
Structured English Immersion Program, 242, 244, 250
Student achievement, 71
Student learning, 76
Student performance, 84
Summative assessment, 79
Summerhill School, 215
Super Bowl, 111, 159
Supreme Court, 91, 248
Survivor, 129, 142
Swift, Jonathan, 270
Systemworld, 115

Tacit Dimension, 176
Tacit knowledge, 90, 176, 177
Taco Bell, 133
TCI, 152
Teacher certification, 6
Teacher deprofessionalization movements, 8
Teacher deskilling, 15
Teacher proof instructional materials, 74
Technical standards, 71, 73–75, 78, 81, 83
Technology, 138, 144
Technology corporations, 8
Teen Newsweek, 147, 159, 161
Terman, Lewis, 171–172, 203
Test scores, 18, 81, 95, 255
Test validity, 82
Testing, 172–173
Texas, 17
The Advertising Council, 149
The Apprentice, 142
The Bell Curve, 177
The Bluest Eye, 236
The Daily Show With Jon Stewart, 145
The Declaration of Independence Road Trip, 149

The History Channel, 149
The Learning Channel, 149
The Magic School Bus, 148–149
The Millionaire Mind, 170
The National Institute on Drug Abuse, 149
The News Hour with Jim Lehrer, 162, 168
The New York Times, 148, 159–160, 168
The Scholastic, 147
The School and Society, 174
The Wall Street Journal, 160
The Weekly Reader group for K-8 readers, 145–147, 163
Thomas, Piri, 269
Thoreau, Henry David, 263
Thurman, Wallace, 269
Time Corporations, 160
Time Warner, 150, 161–163
Tinker v. Des Moines, 125
Title I, 256
Tolstoy, Leo, 193
Tom Jones, 195
Toyota Camry, 266
Toys "R" Us, 151
TQM (Total Quality Management), 119
Tracking, 173
Traditional classroom, 105, 108
Traditional intellectuals, 262, 274
Transition Advisory Committee on Trade, 255
Transitional Bilingual Education, 243
Transmissional pedagogy, 71, 75
Trump, Donald, 129, 143
Turner Learning, 158, 162
TVA (Tennessee Valley Authority), 4

Understanding by Design, 112
Uniform Lists, 91
Unz, Ron, 241–247, 249–250, 252–253, 257–258
UPS, 149
Urban Outfitters, 163–165
U.S. English Foundation, 251
U.S. English Foundation, Inc., 247

Valdez, 131
Validity, 98
Van Manen, Max, 20–21
Varela, Francisco, 225
Viacom, 163
Vietnam, 172

Virgil, 185
Voigt, Cynthia, 269
Voucher supporters, 2
Vygotsky, Lev, 85, 172, 179

Wall Street, 19
Warner Home Video, 148
Webber, Julie, 139
Weekly Reader, 159, 165
Weiler, Kathleen, 131
Weininger, Otto, 200
Wesleyan University, 145
Western Literature, 270
What Do Our 17-Year-Olds Know?, 94, 97
What on Earth, 152
White House, 166
White nationalism, 249
Whittle, Christopher, 155
Whittle Communications, 156–157
Widmeyer-Baker, 19
Wild, Jonathan, 194–195

Winters, Ed, 157
Wolf, Christa, 229
Wollstonecraft, Mary, 200
Women and genius, 199
Women's Movement, 16
World War II, 92, 151, 172
WRC Media Holdings, 146, 147
WRC (The Weekly Reader Corporation), 145, 151, 153, 156, 161
Wright, Richard, 233
WTO, 142

Xavier, Emanuel, 237
Xerox, 145

Yahoo!, 159
Yerkes, Robert, 172
Young, Michael, 177

Zeta-Jones, Catherine, 137
Zone of proximal development (ZPD), 179